Sacred Cells?

Sacred Cells?

Why Christians Should Support Stem Cell Research

Ted Peters,
Karen Lebacqz,
and Gaymon Bennett

ROWMAN & LITTLEFIELD PUBLISHERS, INC.
Lanham • Boulder • New York • Toronto • Plymouth, UK

ROWMAN & LITTLEFIELD PUBLISHERS, INC.

Published in the United States of America
by Rowman & Littlefield Publishers, Inc.
A wholly owned subsidary of The Rowman & Littlefield Publishing Group, Inc.
4501 Forbes Boulevard, Suite 200, Lanham, Maryland 20706
www.rowmanlittlefield.com

Estover Road
Plymouth PL6 7PY
United Kingdom

British Library Cataloguing in Publication Information Available

Library of Congress Cataloging-in-Publication Data

Peters, Ted, 1941–
  Sacred cells? : Why Christians should support stem cell research / Ted Peters,
Karen Lebacqz, and Gaymon Bennett.
       p. cm.
  Includes bibliographical references and index.
  ISBN-13: 978-0-7425-6288-2 (cloth : alk. paper)
  ISBN-10: 0-7425-6288-3 (cloth : alk. paper)
  1. Stem cells—Research—Moral and ethical aspects. I. Lebacqz, Karen, 1945–
II. Bennett, Gaymon, 1972– III. Title.
  QH588.S83P48 2008
  174.2'8—dc22   2007052094

Printed in the United States of America

♾ ™ The paper used in this publication meets the minimum requirements of
American National Standard for Information Sciences—Permanence of Paper
for Printed Library Materials, ANSI/NISO Z39.48-1992.

~

Contents

Preface

Misunderstandings about ethics are like TV commercials and online pop-ups. They are everywhere, it seems. They are pesky. They interrupt. They annoy because they divert attention. Replacing misunderstanding with clarity and focus is the goal of this book.

We start with five misunderstandings. The first is this: *Science is fast, whereas religion is slow; religion is always lagging behind science.* One cartoon from the late 1990s depicted giant black footprints labeled "science" leading off toward the horizon. A little man was pictured running to catch up. He was labeled "ethics." Another cartoon pictured two horses, the faster one escaping from the slower. The faster one was labeled "cloning and stem cell research" and the slower one "ethics."

Sometimes, this is true: Religion chases science and tries to catch up hastily by throwing some ethics at it. But by no means is this universally the case. The stem cell controversy is a good example of a case in which science and ethics worked hand in hand from the beginning.

The three authors of this book are connected with the Center for Theology and the Natural Sciences (CTNS) at the Graduate Theological Union in Berkeley, California. When the worldwide Human Genome Project commenced in 1990, so did CTNS's monitoring of the research in the name of theological and ethical interests. A national team of geneticists, theologians, and philosophers worked on a U.S. National Institutes of Health–funded project, "Theological and Ethical Questions Raised by the Human Genome Initiative."

Nearly two years before the first human embryonic stem cells were isolated and characterized, two CTNS theologians were busy drawing plans for an ethics advisory board to work in tandem with the laboratory scientists. We tell this story here, seeking to convey this point: The full story of stem cell research cannot be told without the chapters on ethics and religion being told with it. In this case, science and ethics coordinate.

The second on our list of misunderstandings is this one: *The stem cell war is one more example of the conflict between science and religion.* Again, this is wrong. Yet, it is widely believed. Take for example an article in one of the world's most respected science journals, *Nature*: "The tension between faith and science never fully subsides. And as these realms regularly come into contact, over everything from Darwin to Dolly the cloned sheep, they sometimes collide with explosive force."[1]

This is wrong for two reasons. First, all relevant religious voices sing praises to the advances of medical science. All encourage science to strive for the betterment of human health and well-being. Some theologians even see scientific research as a divine vocation. No categorical rejection of science exists in mainline Christianity or Judaism, despite the popular image of warfare. Second, in the middle of the ethical debate over stem cell research, some theologians argue for ethical approval of stem cell research. We authors of this book belong in this camp. The disagreements over stem cell research are not due to a conflict between science and faith. Rather, they are due to honest yet differing interpretations of what faith requires at the present moment. While some religious people may oppose specific scientific advances, most welcome and support them. We are among those people of faith who welcome and support stem cell research.

The third widespread misunderstanding is this: *The principal job of the ethicist is to say "no" whenever possible.* As ethicists, we are often asked *ad nauseam*, "Where do you draw the lines?" We presume that "lines" mean fences, and scientists should not jump fences. Not only does ethics *chase* science, but its role is to *chastise* science. The presumption is that to be an ethicist is to put fences around scientists. Where did this image come from?

To our chagrin, the laboratory scientists with whom we work complain all too often that their contacts with ethicists are routinely negative. The job of the ethicist is to put up a "no trespassing" sign, so it seems to them. When such scientists see an ethicist coming, they run to hide. They are understandably reluctant to engage us in conversation.

The three coauthors of this book have asked ourselves: "Is this really what we as ethicists want to do? Is it our job to say 'no'?" Certainly not. The job of the ethicist is to display with as much rigor as possible the important ethical issues at stake and to work with people in making ethical decisions. Sometimes this *does* mean saying "no." But just as often, it means saying "yes." We conceive of ethics as providing helpful guidance to scientists and others faced with difficult problems and oriented toward a better future.

The fourth and related misunderstanding presumes that *every religious ethicist says "no" to stem cell research*. The job of the religious ethicist, it seems to many, is to describe stem cell scientists as baby killers who are cannibalizing early embryos to make spare body parts. If you are a religious ethicist, it is assumed that the only question you ask is: What is the moral status of the embryo? However, we believe other important questions need to be asked as well, especially this one: How can medical science improve human health and well-being? In the case of stem cell research, the potential for a dramatic leap to increased human health is significant. In our considered judgment, saying "no" to stem cell research would be immoral. We seek to provide ethical justification for a positive affirmation of this particular line of scientific research.

This book will be both descriptive and prescriptive. First, we describe the worldwide stem cell debate in terms of competing ethical frameworks. We hope this will illuminate the debate and make the apparent impasse more understandable. Then, in addition, we provide the reader with our own prescription, namely, that religiously minded persons should support research leading to stem cell therapies.

We intend to show that public policy warfare regarding stem cell research is a cross fire coming from multiple directions. There are in fact multiple frameworks available for addressing ethical issues. One of these just mentioned is the *embryo protection framework*. This is the framework within which many religious people formulate their positions and accuse the scientists of promoting a "culture of death." Scientists respond by saying that the preimplantation embryo is not a human person, even *in potentia*. The central question—the one that currently dominates the public debate—is whether the early embryo possesses morally protectable dignity, so that destruction for purposes of research is forbidden. Is stem cell research akin to abortion? One side answers "yes." The other side answers "no." Both sides tacitly agree that the rightness or wrongness of stem cell research will be determined by the moral status of the early embryo.

Yet, this is not the only possible framework. It is a mistake to think that this is the only way in which the moral battle can be waged. At least two other frameworks appear on the moral map.[2] One of these we label the *human protection framework*. The essential question here is: How can we protect our humanity from the hubris of science, technology, and other human ventures?

What is central in this framework is the protection of a sense of what is "essential" to human life and dignity, perhaps even a reverence for what is natural in making us the human beings that we are. Human nature protectionists are driven by anxiety expressed in Aldous Huxley's novel of the early 1930s, *Brave New World*, when genetics was becoming a household word. The words "brave new world" connote a scientized and technologized civilization in which the biological sciences have placed the human race under totalitarian control. Today, whenever it appears that scientists are manipulating something internal to human nature—something we deem essential, such as DNA—the specter of the brave new world arises. Is DNA sacred, so that we should ask our scientists to keep their hands off it? Or, is DNA simply one more resource for research leading to genetically engineered improvements in medical care and human well-being?

Nobody wants to create a brave new world. The question is whether our true humanity is to be found in nature alone apart from modification by technology, or is it found in self-improvement through the science and technology we human beings have created?

The third framework is the *future wholeness framework*. Here the attention is given to the dramatic potential of regenerative medicine. What human embryonic stem cell research is leading to is a quantum leap well beyond any previous form of medical therapy. It is leading toward the actual regeneration of organs such as the heart, liver, pancreas, and even the brain. If we could do more than merely stop deterioration—if we could actually cause new tissue to grow to replace tissue damaged by disease or accident—recipients of stem cell therapy could emerge healthier and stronger than they previously were. Even though still in the theory stage with animal studies and some clinical studies, stem cell research shows promise not just for amelioration but for actual cures for many types of cancer, Parkinson's disease, heart disease, diabetes, Alzheimer's disease, and many others. According to the future wholeness framework, the promotion or blocking of such research is itself a moral issue. When such potential for relief of suffering and betterment of human life is judged to be a realistic potential, then a moral obligation to pursue it kicks in. Within this framework, arguments to shut down such research require considerable burden of proof.

From within the future wholeness framework, we ask: Is it moral for religious advocacy groups to shut down research that could lead to relieving the suffering of millions if not billions of persons in the future? We also ask questions related to justice: Recognizing that the advance of stem cell knowledge will be staggeringly expensive, how will the medical products be distributed? Will the poor persons of the world have access to the marvels of this science?

How can we structure the economics of the medical delivery system so that benefits are distributed worldwide? Such ethical questions get ignored when we presume that the question of embryo protection is the only question on the ethical agenda.

The three authors of this book place ourselves primarily within the third framework. As people of faith, our ethical commitment begins with a sense of God's promise for an abundant future and therefore with a commitment to improve the human lot in life. We believe support for medical research is support for improving human health and well-being. Regenerative medicine could lead to a much more abundant life for many among us. Having made this commitment to work primarily within the future wholeness framework, however, we still feel obligated to engage our friends and opponents within all three ethical frameworks, and we will do that in subsequent chapters.

After weighing the arguments in all three frameworks, it is our considered judgment that stem cell research should go forward. It is our further recommendation that public policy support such research on behalf of the welfare of all and on behalf of future generations who will benefit from the advance of medical science in this generation. We contend that religious believers of our own persuasion and the faithful of other traditions should hold such science in high regard and pray for its success. As Christian theologians, we say "yes" to stem cell research. In what follows we will explain why.

This leads to the fifth on our list of widespread misunderstandings: *Supporters of stem cell research are pro-choice on abortion and generally disregard the human right to life. They violate the sacredness of life.* This misunderstanding is purveyed widely in the media when oversimplifying its reports on coalitions of Christian groups protesting alleged human rights violations. In order to divide our populace into neat factions, the media tells us that those who want to pull the feeding tube from a person in a perpetual vegetative state are also pro-choice and favor stem cell research. The news is lumpy—that is, it lumps otherwise disparate causes together as if this is accurate and informative.

We plan to show what a big mistake it is to lump together the abortion controversy with the stem cell controversy. Even though there is some overlap, these two issues have significant ethical differences. Therefore, the position one takes on one issue does not dictate how one will deal with the other issue. It is quite possible to favor stem cell research and oppose elective abortion. In fact, one of the authors of this volume would fall on what is usually called the "pro-life" side of the ledger regarding elective abortion, while the other two take the pro-choice position. The two pro-choice supporters report, "This is *because* we're pro-life." All three of us feel strongly committed to human rights; and we believe this strong commitment derives from the fundamentals of our shared faith. Yet, we contend that the moral logic of stem cells is different from the moral logic surrounding abortion. One cannot simply lump them together under the guise of protecting the "sacredness" of life.

One of the difficulties with those who wish to equate the abortion controversy with the stem cell controversy is that they tacitly treat DNA as sacred; they treat cells as if they were persons. To treat anything as sacred means to treat it as something that cannot be violated, to treat it as a source of moral value. Tacitly, embryo protectionists treat prepersonal cells as sacred, protecting the dignity of the stem cell as if it were a person. We believe that God and God alone is sacred; and we believe that human persons should be treated with dignity. To clarify what we mean here is one of the tasks of this book.

The three of us coauthoring this book are theologians with a special interest in ethics, especially bioethics. We have studied the issues surrounding human embryonic stem cells. In fact, we have been present while the very plan for isolating these cells was being conceived and the initial discoveries made. We have examined the arguments put forth by religious leaders and others who want to shut down stem cell research. We have reviewed them carefully and respectfully. It is our considered judgment that decisive moral arguments can be lifted up in support of scientists engaged in this research, chief of which is this: *Medical research into the regenerative potential of human embryonic stem cells fulfills the principle of beneficence*—that is, it fulfills our divine mandate to improve human health and well-being—and does not violate other important principles. As theological ethicists, we want to say "yes" to stem cells.

We are all Protestants, but from different traditions. Indeed, we cover the spectrum from liberal to conservative traditions within Protestant theology. As scholars, we try to explicate the views of every perspective with accuracy, sympathy, and fairness. As ethicists, we feel obligated to render judgments and support these judgments with sound reasoning based upon our deeply held Christian commitments. Although we understand the unavoidable complications and nuances of the public policy debates over genetic research, we can say, with some qualification: "yes" to stem cell research. The book that follows will say why.

Ted Peters
Karen Lebacqz
Gaymon Bennett

Notes

1. Tony Reichhardt, "Studies of Faith," *Nature* 432 (2004): 666.

2. Readers following this debate will notice in this book a change in vocabulary in describing these frameworks. In previous articles and in the book by Ted Peters, *The Stem Cell Debate* (Fortress Press, 2007), the second framework was called the "nature protection framework;" we now call it the "human protection framework." What was previously designated the "medical benefits framework" we now call the "future wholeness framework." The reasons for these changes will become apparent in later chapters.

CHAPTER ONE

~

The Ethical Prehistory of Stem Cells

Significant events have a prehistory and a posthistory. In 1981 the mouse embryonic stem (ES) cell was discovered by scientists in Great Britain and in the United States. This discovery led to the exciting possibility that there might be an equivalent cell in the human being. An *in vitro* fertilization (IVF) clinician in Singapore, Ariff Bongso, first located human embryonic stem (hES) cells in culture but was unable to get them to replicate indefinitely. Finally, James Thomson at the University of Wisconsin, the first to find ES cells in monkeys, had the major breakthrough in 1998. Thomson isolated hES cells, and with this the worldwide controversy over stem cell ethics exploded. Or, so it seems.

This scientific prehistory of hES is well known, at least to scientists. What is less well known is that there is an ethical prehistory as well. It is commonly thought that interest in the ethics of hES cell science followed the announcement of Thomson's isolation of hES cells in 1998. But that is not so. Controversy—or at least, discussion—preceded this historic announcement.

On a spring day in Berkeley, California, in 1996, a group of theologians and scientists gathered in the Dinner Board Room of the Flora Lamson Hewlett Library at the Graduate Theological Union (GTU). This historic meeting not only represents the initiation of formal ethical discussions of stem cell research in the United States, but it also represents a new chapter in the story of the relationship between biotech research scientists and ethical advisory boards. Like streams from separate mountain glaciers first trickling and then

1

converging into a single flowing river, basic scientific research converged with theological ethics to create a new flow of public policy discussion.

The SyStemix Stream

The first prehistory stream comes from ethical questions arising within the research program at SyStemix Corporation in Menlo Park, California. Founded in 1988 by Stanford University immunologist Irving Weissmann, SyStemix sought nothing short of curing AIDS, other autoimmune diseases, and cancer. No vaccine could work for AIDS, concluded SyStemix's scientists, because the HIV virus mutates every 36 hours. It would not be possible to continue revising the vaccine formula or antivirals to keep up with such rapid change. An alternative strategy would be needed. Instead of eliminating HIV, asked the researchers, could the patient continue to live with HIV that does not progress to AIDS? If the bone marrow could continue to produce new and AIDS-resistant blood—healthy blood with immune cells that could not be infected, having been rendered resistant via genetic engineering of the hematopoietic stem cell—this would be possible.

Weissmann had previously isolated hematopoietic (blood making) stem cells in mice. At SyStemix, scientists turned their attention to repeating this, to trying to create a human cell equivalent. One in every 300,000 blood cells is a stem cell. Could the stem cells be extracted from the blood, concentrated, and injected into bone marrow, where they would continue to generate new and healthy blood despite adverse conditions? The answer turns out to be yes.

This was made possible by SyStemix's use of an invention by Mike Mc-Cune, a cofounder of SyStemix. McCune had been a postdoctoral fellow in Weissman's Stanford University lab. SyStemix at its founding had licensed this invention from Stanford. The invention involved the use of the severe combined immunodeficient (SCID) mouse, a mouse without its own functioning immune system, that could therefore accept organ fragments from any species, including the human species, without rejection. When receiving human tissue without rejection, it became known as the SCID-Hu mouse. SCID-Hu made the right kind of experimentation possible.

SyStemix scientists also invented a process for harvesting and concentrating human hematopoietic stem cells and patented it in 1990. U.S. Patent number 5,061,620 covers both the method for obtaining these stem cells and for the resulting composition, the product.

In 1990 Linda Sonntag became CEO of SyStemix. By 1992 she perceived ethical dark clouds forming on the stem cell horizon. Sonntag understood the interface between research and ethics. Back in 1986 while at Focus Technologies in Washington, D.C., she had convened a six-person panel to sort through ethical issues surrounding ownership of patient information and protection of individuals from genetic discrimination. At SyStemix in 1992 Sonntag made arrangements for two theologians working in bioethics at the GTU in Berkeley, Karen Lebacqz and Ted Peters, to visit SyStemix and discuss emerging dilemmas.

The Graduate Theological Union Stream

How did Sonntag know that resources could be tapped in Berkeley? Sonntag was a friend of one of Karen Lebacqz's doctoral students, Suzanne Holland, who at this writing teaches ethics at the University of the Puget Sound.[1] Lebacqz was then professor of ethics at Pacific School of Religion, a member seminary of the GTU.

The GTU is a consortium of theological seminaries offering a fully ecumenical and interreligious context for theological education. In addition to seminaries, the GTU also houses research centers, which include faculty both from its partner institute and neighbor, the University of California, Berkeley. One of these research centers, the Center for Theology and the Natural Sciences (CTNS), played a key role in the ethical prehistory. When Nobel Prize winner James Watson in 1987 pressed the U.S. Congress to support what would later come to be known as the "Human Genome Project," he advocated that 5 percent of the government's budget be devoted to Ethical, Legal, and Social Implications (ELSI) of genetic research. As soon as Congress passed legislation and the National Institutes of Health (NIH) established the National Center for Human Genome Research, CTNS applied for and received an ELSI grant. Ted Peters served as principal investigator for "Theological and Ethical Questions Raised by the Human Genome Project." Professor Lebacqz served as a member of CTNS's core research team, and doctoral student Suzanne Holland worked as a research assistant.[2]

It was Suzanne Holland who mediated the connection between Linda Sonntag and the two GTU professors, Karen Lebacqz and Ted Peters. The GTU stream was about to flow into the SyStemix stream, and eventually into the surging river of stem cell controversy.

The SyStemix and GTU Confluence

Sonntag presented three ethical problems to Lebacqz and Peters. The first on Sonntag's list was the use of fetal organs in a humanized mouse.

As mentioned above, SyStemix was using the SCID-Hu mouse for all of its research efforts. These SCID-Hu mice were constructs of SCID mice that were then implanted with fragments of human fetal organs, tissue derived from aborted fetuses. SyStemix was concerned about potential concerns regarding the use of abortuses for experimental purposes. The use of fetal tissue in other forms of research had already become a matter of international concern. Many ethicists could easily distinguish the decision to legalize abortion from the decision to utilize the resulting abortus as a research subject. SyStemix was not responsible for the legalization of abortion, nor was it responsible for individual women making the decision to terminate pregnancy. SyStemix would not be morally culpable in this respect. Once it had been determined scientifically that fetal tissue provided unusually valuable material for their medical experiments, they proceeded to utilize this material. If the field of ethics can be considered a helping field—that is, helping persons or societies to work through real-life moral dilemmas—then what needed to be asked here is this: In light of the existing circumstances, what is the best way forward? It was our judgment that such research should go forward.

This ethical problem confronted by SyStemix in 1992 foreshadowed one that would arise again six years later regarding hES cells. One concern in the rising controversy was the destruction of the early embryo in order to establish ES cells. Instead of aborted fetuses, however, the new wave of researchers would be using human zygotes produced through IVF and discarded by fertility clinics. Again, the researchers would not be responsible for producing the research material. The question would become: Is it morally licit to use this material for medical purposes?

The second of the three ethical problems on the Sonntag list had to do with a side implication of the successful isolation of human hematopoietic stem cells. The ability to inject such cells into the bone marrow and to guarantee continued production of healthy blood placed SyStemix on the brink of providing a decisively effective therapy for leukemia. Word having gotten out about the discoveries of SyStemix resulted in a line at the door of desperate leukemia patients seeking stem cell injections. Lebacqz and Peters momentarily celebrated the medical achievement, congratulating Sonntag. "So, what's the problem?" they asked.

"We are not ready for clinical application," Sonntag stressed. "We want to stick to our research mandate. If we were to allow one of these leukemia patients to use our concentrations of blood stem cells, and should something go wrong, and should they decide to file a lawsuit, then SyStemix would be financially crippled and we could no longer pursue our long-range goals."

"Yes," acknowledged Lebacqz and Peters, "this is a problem." How does one weigh the alleviation of immediate human hardship against long-range goals? Pressures to proceed quickly to clinical application always arise when new technologies appear to offer hope, but it is important that movement not jeopardize patient health or eventual success of the therapy.

The third ethical problem was introduced in 1992 but became a widespread public controversy in 1995. It has to do with the legitimacy of the SyStemix composition patent. In 1993 Andrew Kimbrell published a book *The Human Body Shop*, in which he describes an unnamed California company that allegedly patented human body parts. "In 1991 the Patent and Trademark Office (PTO) granted patent rights to a California company for commercial ownership of human bone marrow 'stem cells' (stem cells are the progenitors of all types of cells in the blood). The PTO had never before allowed a patent on an unaltered part of the human body."[3]

This might have gone relatively unnoticed by the reading public if it were not for the author's colleague, Jeremy Rifkin at the Foundation on Economic Trends in Washington. On May 18, 1995, Rifkin convinced 180 religious leaders to sign a statement supporting his strong stand against the alleged patenting of the human genome. At a press conference Rifkin called for a government ban against patenting genes and genetically engineered animals. One of the religious leaders, Richard Land of the Southern Baptist Convention, was quoted in the *New York Times* saying, "Instead of whole persons being marched in shackles to the market block, human cell lines and gene sequences are labeled, patented and sold to the highest bidders."[4] Both the biotech industry and the U.S. government were rhetorically indicted for crass disavowal of the sacredness of human life and for cannibalizing human bodies for spare parts in order to make a profit. Eventually attention turned in the direction of SyStemix. In a phone conversation with Peters, exasperated Sonntag expressed frustration that misunderstanding could be so widespread.

A sober review of what was happening will show that the accusations against SyStemix or the government patent office were unfounded. Nothing like body parts had been patented. What the U.S. Patent and Trademark Office has required since the days of Ben Franklin is that inventions exhibit three qualities: they must be novel, nonobvious, and useful. The term "body parts" suggests that a concentration of hematopoietic stem cells already occurs in nature and cannot be considered as the novel invention of a human artificer. However, the PTO concluded that such a concentration does not in fact occur in nature; rather, it took a sophisticated process and a machine to make purification and concentration possible.

One could appeal to the vitamin B-12 pill as a precedent. Although vitamin B-12 appears in minute quantities in the livers of cattle and in certain other microorganisms, it takes human technology to concentrate it and make a convenient pill. The purification process makes concentrated vitamin B-12 novel. Hence, the PTO felt justified in granting a patent for it, and this was sustained in the courts. This also was the case with hematopoietic stem cells in purified and concentrated form.

Our ethical deliberations and analysis of this problem concurred that the SyStemix patent fits squarely within the two centuries of patent tradition and criteria. Accusations that human body parts or someone's private blood cells were becoming the property of a for-profit company appeared to be misguided. Lebacqz and Peters sought to distinguish ethical deliberation from political rhetoric.[5] With the confluence of the SyStemix and GTU streams, the ethical prehistory of the stem cell controversy was almost complete.

The Geron Stream

Enter the Geron Corporation. Michael D. West, founder of the Geron Corporation in Menlo Park, set out to isolate and characterize hES cells. Whereas hematapoietic stem cells are only multipotent—that is, able to generate all cell types within the blood stream—ES cells are pluripotent, making them capable of producing any and every tissue in the body. ES cells are the progenitors of hematopoietic stem cells, which are the progenitors, in turn, of new blood cells. West had previously worked with ES cells in mice. He began anticipating repeating this with human beings.

West, holding a Ph.D. from Baylor College of Medicine, founded the Geron Corporation in 1990, and until 1998 he initiated and managed programs in telomerase diagnostics and therapy. West selected the name "Geron," which in Greek means "old man." It comes from the New Testament, from the passage where an old man named Nicodemus asks Jesus, "How can a man be born again when he is old (geron)?" (John 3:4).[6] The name "Geron" has morphed into our modern word, gerontology. West's company had a primary interest in research that might extend the life span and reduce the impact of aging.

West was passionate about his vocation as a scientist and audacious about his research goal, to say the least. The goal he set for himself was nothing less than the defeat of death. West describes the moment when he realized his life's calling: "It was crystal clear to me what I had to do. I had to defeat death. . . . This was inarguably the greatest and highest calling of mankind,

to find and control the biological basis of the immortality of life, and to alleviate the suffering of our fellow human beings."[7] West wanted to know if
science could win the battle against the Grim Reaper.

The first skirmish won in this battle was the discovery of the time clock
that ticks away through cell division, deterioration, and demise. The clock
ticks with the shortening of our telomeres. A telomere is a sequence of
nucleotides on the ends of each chromosome. Literally, it is the following
sequence: TTAGGG.[8] To halt deterioration of the telomeres and make the
clock tick longer became the first goal of Geron. This ability was achieved
during Geron's first half-decade.

Geron's telomerase research built upon the foundational work of Aus-
tralian born molecular biologist, Elizabeth Blackburn, who along with her
colleague at the University of California at San Francisco, Carol Greider,
had done the pains taking work of sequencing the telomeres. Geron brought
Blackburn in as a consultant in order to further this research.[9]

West was then asking: Might we find a form of human cell whose telomeres never shortened, cells that would be effectively immortal? With immortal cells, could we then regenerate tissue damaged by disease or trauma?
Even if science cannot help Nicodemus or the rest of us to become born
again, could it help our bodily organs become born again?

West turned to stem cells. If we could isolate stem cells, might we gain
the ability to regenerate tissues or even organs, thus overcoming some of the
most difficult diseases of aging, such as heart disease? The stem cell became
the target for research because of two virtues: if immortal it could have the
power to regenerate tissue, and if pluripotent, it could be directed into making any tissue we designate. What do we mean by this?

Immortality?

First, *immortality*.[10] With this somewhat surprising word borrowed from theology and imported into science, West meant a cell that would continue
to divide and divide without deterioration, in principle, forever. What he
had learned from his previous research into telomeres is that, as long as the
telomeres are intact, a cell would continue to remain healthy and generate
new tissue. The achievement of the Geron scientists was that they found a
way to fire a gene producing the telomerase enzyme into a cell that would
lengthen the telomeres or prevent deterioration. Telomerase positive cells
had been created.

This concept of cell immortality has a most dramatic shadow side. When
we turn to cancer cells, the problem is that too much telomerase activity

spawns the unlimited growth of a tumor. The tumor grows and grows until it kills the patient. This observation led Geron scientists initially to look for a way to turn off the telomerase gene. If we could turn off the production of telomerase within a cancer cell, the cell would no longer divide. Eventually it would become senescent (die); the tumor would shrink in size and the patient would recover. The devising of a method for firing a knock-out gene into a cancer cell to turn off telomerase activity was one of the achieved goals of this research. As we write, Geron is now exploring ways of turning this technology into a therapy to fight cancer.

Thus, telomerase activity needs to be shut off in the case of cancer; but it needs to be turned on in the case of regenerative medicine.[11] When turned on for purposes of tissue regeneration, the patient needs to be protected from the possibility of runaway telomerase activity and the creation of a cancer. West set this as one of his goals. By looking for a naturally immortal cell, he believed he could gain access to regenerative power without the risk of precipitating cancer.

Pluripotency?

Second, pluripotency. What West would need would be a stem cell that would be at least pluripotent. What do we mean by this? A stem cell has two important qualities. First, it is clonal—that is, it can replicate or clone itself. It is self-renewing. Second, it produces daughter cells for different types of tissue. *Potency* refers to this ability to generate different cell types.

The potency of cells can be ranked. A cell is *totipotent* (totally potent) when it can make any tissue in the human body and also, under the right conditions such as existing within a woman's body, make a baby—that is, proceed through the stages of embryo development and become a human being. The next level down would be a cell that is *pluripotent*—that is, a pluripotent cell could differentiate into any tissue in the body, even if it is unable to become an embryo. Totipotency, or at minimum pluripotency, is what West was seeking; because he would need cells that could be guided into becoming any tissue or organ he might select: heart, liver, pancreas, brain, and so forth.

When West began, he could have utilized Linda Sonntag's *multipotent* blood stem cells, those affecting different forms that blood take. But this would be insufficient. Virtually useless would be *unipotent* stem cells, those that renew only one form of tissue, such as skin or hair. Unipotent cells are best called progenitors, because they derive from stem cells yet can generate new cells of their own type. West needed more versatility than what a multipotent or unipotent cell can deliver. With all this in mind, West asked: Might he find both immortality and pluripotency in hES cells?

What, then, is a stem cell? It is a cell that is able to reproduce itself throughout the life span of an animal or person; and it will give rise to differentiated somatic cells or other stem cells, or perhaps both. The daughter cells of stem cells may be either differentiated cells or more stem cells.[12] When giving rise perpetually to more stem cells that remain healthy and do not deteriorate, the stem cells are immortal. Could such a treasure be found?

The stem cell treasure would include more than the golden egg. By regenerating new tissue, implanted stem cells in a patient following a heart attack would so strengthen the heart that it would be stronger than it had been before the attack. By teasing stem cells into becoming brain tissue we could develop therapies to overcome Alzheimer's and Parkinson's. By teasing stem cells into becoming pancreas tissue we could overcome diabetes. By teasing stem cells into becoming spinal nerve cells we could repair injuries and overcome paralysis of the lower limbs. And all of this in addition to its potential for winning the war against cancer. The therapeutic potential of regenerative medicine appears like a cornucopia of healing and human betterment.

Pursuing Venture Capital through Ethics

In 1996 West went looking for venture capital to support his endeavors in a spin-off company to be named "Primordia." His two research compatriots in this venture would be Jeryl Hilleman and Andrea Bodnar. One of the first people West spoke with was Linda Sonntag, who by then had left her position as CEO of SyStemix and had become an independent venture capitalist. The prospects of research into ES cells intrigued Sonntag. In her own mind, however, Sonntag felt she needed to think through her fiduciary responsibility to potential investors. Could such research make a profit? Was it ethical?

Now the memories of Sonntag and West differ slightly. As Sonntag recalls it, she introduced West to the ethical implications of his research proposal. In order to press her concern, Sonntag firmly stated that she would not engage in raising money until a full "ethical analysis" had been completed. Sonntag then invited West to a "round table" discussion of ethical issues. West accepted.

According to West, Sonntag did not introduce him to the ethical issues for the first time. He had previously anticipated them. He was educated in an evangelical Christian environment that had sensitized him to ethical issues.

And he had worked through the question of the moral status of the preim-plantation embryo by taking the developmentalist position—the position that the moral status of the embryo increases as the embryo develops. West does agree that Sonntag told him in private, "It will never fly. Everyone will see it as unethical." West reports that he responded by defending his already established moral judgment, namely, that the developmentalist position on the moral status of the embryo would suffice.

The two memories agree about what happened next: Sonntag invited West to a roundtable discussion with theologians in Berkeley at the GTU. West welcomed the idea. Yet he reports that one of his research colleagues, Jeryl Hilleman, was "horrified" at having to speak with theologians. Still, the three—West, Hilleman, and Bodnar—consented.

Sonntag then contacted Suzanne Holland, who in turn asked Karen Leb-acqz and Ted Peters to help set up the roundtable at the GTU. On an April day in 1996, 18 GTU faculty members and graduate students gathered in the Dinner Board Room of the library to greet the 3 representatives from the Geron Corporation. The faculty group included, among others, Robert John Russell from the Center for Theology and the Natural Sciences, William R. O'Neill, S.J., from the Jesuit School of Theology at Berkeley, Richard Gula from the Franciscan School of Theology, and Michael Mendiola from the Pacific School of Religion.

Linda Sonntag directed the meeting. She pointed to Dr. West and said to the assembled group, "Not until we get an ethical analysis that I can live with, will I try to raise the money for human embryonic stem cell research." West then assumed the position of lecturer and spent nearly an hour de-scribing the research protocol he was planning to begin. He said he would create an "immortal line" of stem cells that could make any tissue in the human body and potentially regenerate any diseased or damaged organ. The scientific vision was awe inspiring, even to those outside science who were learning of this for the first time.

When it became clear that isolating hES cells would include the destruction of the blastocyst—the embryo between four and six days after activation—Professor Lebacqz announced, "This is an important ethical issue!"

Before the day was finished, the theologians identified what they con-sidered to be the core ethical issues raised by ES cell research: (1) the moral status of the embryo; (2) the future unforeseen consequences of these therapeutic interventions; (3) the revolutionary therapeutic potential of the research; and (4) the economic justice concern for worldwide distribution of the medical benefits. These ethical issues would eventually explode into the global stem cell debate. In April 1996, more than two years before scientists were able to isolate hES cells, the global controversy over the ethics of ES cell research had quietly begun.

The first of these four, the moral status of the embryo, received part but not all of the attention at the roundtable discussion that day. The Roman Catholic theologians pressed for further elaboration. In the process, the Catholics affirmed their Church's commitment to protecting the life of the early embryo. Even in the face of the prospect that such laboratory research could provide untold medical benefits, the Catholics could not in good conscience countenance the destruction of potential human beings in the service of health and medicine.

Sonntag listened carefully. She reports hearing what she discerned as "the sense of abhorrence in the Catholic reaction and an ethical stance with no apparent wiggle room." If this Catholic reaction should become a general public reaction, she thought to herself, then reception for Geron's or Primordia's program would be undercut. The funding would be undercut, and so would the market. On this day, Sonntag decided to decline West's request to raise capital to support the research. Geron would have to look elsewhere for its money.

Notes

1. Suzanne Holland, Karen Labacqz, and Laurie Zoloth, eds., *The Human Embryonic Stem Cell Debate: Science, Ethics, and Public Policy* (Cambridge, MA: MIT Press, 2001).

2. Ted Peters, ed., *Genetics: Issues of Social Justice* (Cleveland, OH: Pilgrim Press, 1998).

3. Andrew Kimbrell, *The Human Body Shop* (San Francisco, CA: Harper, 1993), 210.

4. "Religious Leaders Prepare to Fight Patents on Genes," *New York Times*, May 13, 1995, Front page, National Edition.

5. See Ted Peters, *Playing God? Genetic Determinism and Human Freedom* (New York: Routledge, 2nd ed., 2002), 120–22.

6. Michael D. West, *The Immortal Cell: One Scientist's Quest to Solve the Mystery of Human Aging* (New York: Random House/Doubleday, 2003), 90.

7. West, *Immortal Cell*, 30.

8. "The word 'telomere' comes from the ancient word *telos*, meaning 'end,' and *mere* meaning 'parts.'" West, *Immortal Cell*, 61. For a discussion of TTAGGG, see pages 79–81, 100. When the telomere becomes too short, the frayed ends of the chromosome attempt to fuse with another chromosome. It appears that the critical length is 12.8 repeats or 6 base pairs; any shorter than this and the loosened chromosome end may begin the fusion process. Heidi Ledford, "Minimum Telomere Length Defined for Healthy Cells," *Nature* 449 (2007): 515.

9. Catherine Brady, *Elizabeth Blackburn and the Story of Telomeres* (Cambridge, MA: MIT Press, 2007), 149–50.

10. Since 1996, the scientific vocabulary has shifted somewhat. "Immortal" is used less frequently. Certainly stem cells proliferate; yet "terms such as 'immortal' and 'unlimited' are probably best used sparingly if at all," at least with reference to adult stem cells. This according to Douglas A. Melton and Chad Cowan, "'Stemness': Definitions, Criteria, and Standards," in *Handbook on Stem Cells*, 2 volumes, ed. Robert Lanza (Amsterdam: Elsevier Academic Press, 2004), II: xxiv.

11. The telomerase theory is still undergoing development, and it has its critics. Stanley Shostak, for example, points out that "the roles of telomeres in cell senility and carcinogenesis seem contradictory. Telomeres should be longer in cells that divide indefinitely," but in fact telomeres in cancer cells are often shorter than their normal tissue counterparts. Stanley Shostak, *Becoming Immortal: Combining Cloning and Stem-Cell Therapy* (Albany: SUNY, 2002), 27. It appears that the length of the telomere in itself does not determine the level of cell activity.

12. See Shostak, *Becoming Immortal*, 177.

~

CHAPTER TWO

From Science to Ethics in a Flash

Despite the parting of the ways between West and Sonntag, West had become convinced that bringing ethicists into the research at an early point would be very worthwhile. In fact, professional ethical consultation would be indispensable. Because of the Graduate Theological Union (GTU) meeting, West reports that he "hit on the idea of an ethics advisory board." The Geron Corporation would establish an "Ethics Advisory Board" (EAB) to reflect on the scientific research as it unfolded.

West asked for a follow-up meeting with Lebacqz and Peters, which took place a few weeks later over pea soup and sandwiches at the Stuffed Inn just north of the University of California campus in Berkeley. These three laid the plans for what would become the Geron EAB. Later, West wrote the role to be played by his envisioned EAB: "Private companies have no legal obligation to set up such a committee, but the standard I wanted to apply was that every move we made should be absolutely beyond reproach."[1]

The board should be small, they said, so that it could deliberate thoroughly. Even if small, it should be representative. As the three talked, they deliberated aloud. The proposed EAB should include, if possible, representation from Jewish, Protestant, Catholic, and secular perspectives. Should it also include Hindu, Buddhist, Islamic, or other representation? Would so many persons make it unwieldy? Despite the desired diversity, the planners agreed, the EAB should be sufficiently harmonious so as to engage in progressive cooperative research on its own. At the conclusion of the Stuffed Inn planning session, it appeared that the formation of the EAB was imminent.

In West's mind, the EAB would have "veto power" over research protocols deemed unethical. When West got back to the office with his idea, however, the corporate authorities at Geron rejected the ethical veto. This would violate their corporate fiduciary duty, they contended. An implementation delay set in. The idea of the EAB went into stall.

The delay lasted 28 months. In the meantime, other things were happening. In the summer of 1996 Geron went public, and the management became worried about pulling off the IPO successfully. Geron decided to keep the stem cell operation, making it impossible for West to fund the nascent Primordia. West decided to leave Geron in February 1998. Eventually West moved to Boston to start a competitor company, Advanced Cell Technology (ACT). In Boston, West convened his own ethics board, which has included some of the leading bioethicists in North America, such as Ronald Green.

Meanwhile, back at Geron, the stem cell agenda was pressed forward by signing research contracts with, among others, the University of Wisconsin and Johns Hopkins University. James Thomson at the University of Wisconsin, with Geron funds, is credited with the first isolation of human embryonic stem cells in August 1998.[2] (As noted earlier, a Singapore team led by Ariff Bongso had already isolated cells from a human blastocyst in 1994; but it was Thomson who characterized or "immortalized" these cells in 1998.[3]) A month later, John Gearhart, also using Geron funds, first isolated human embryonic germ (called hEG) cells, a near equivalent to embryonic stem (ES) cells.[4]

In the contract with Johns Hopkins dated August 1, 1997, the university stipulated that Geron must create its own ethics advisory board. The clause in the contract was actually put there by Michael West during the contract negotiations. West had thought to himself, "If someday I should leave Geron, then Geron would be required to keep the Ethics Advisory Board."

Thomas Okarma, then director of research, dusted off the shelved EAB files and followed up by contacting Lebacqz and Peters. He conducted his own series of interviews and selected an EAB membership that looked quite like the one earlier planned with West. The first meeting was convened in July 1998 with a membership including Michael Mendiola (Roman Catholic), Laurie Zoloth (Jewish), Ernle W. D. Young (secular), along with Lebacqz and Peters (Protestants). In the months that would follow, Gaymon Bennett was added as the research assistant. When Mendiola left the board, he was replaced by veteran Catholic bioethicist Albert Jonsen. The board continued in a quarterly consultative capacity until disbanded by Geron in 2002.

First the EAB, Then the Stem Cells

The first Geron EAB meeting was held July 7, 1998. At that meeting, Karen Lebacqz was elected as the first chair of the board, with the expectation that the position would rotate after several years. The first meeting was primarily a "fact-finding" meeting, learning about the stem cell research that was currently sponsored by Geron. The board set about developing a list of criteria for ethical research into stem cell technologies.

That first meeting was scientific boot camp, basic training. But by the second meeting, training gave way to active involvement. Okarma called the EAB members for an urgent second gathering in early September. Once the group was assembled, he announced that James Thomson at the University of Wisconsin had successfully isolated human embryonic stem (hES) cells by creating four characterized lines. Three of these lines were derived from thawed and activated frozen embryos from an *in vitro* fertilization (IVF) clinic; one was not frozen. Thomson had pursued his research with Geron funding. When Okarma presented the details of Thomson's research, Lebacqz and Peters could see that this was just what Michael West had been planning.

A month later, John Gearhart at Johns Hopkins successfully isolated hEG cells. The difference is this: Whereas the hES cells are derived from the embryo at the blastocyst stage, hEG cells are derived from aborted fetuses between five to eight weeks after conception. Both contain 46 chromosomes. Both sources would be totipotent. Geron had been funding both projects, two separate possible routes to the same destination.

Okarma stressed absolute confidentiality. "Don't even tell your spouses," he emphasized. The publication of this research would not take place until October or November. In the meantime, the Securities and Exchange Commission would be watching for any insider influence on stock trade. The EAB was admonished not to purchase any Geron stock. Okarma wanted to eliminate all risk of insider trading and conflict of interest. All conversation outside Geron doors was embargoed for three months while the secret was kept.

The EAB set about writing up its evaluation of the research and devising a list of ethical issues. Lebacqz chaired a drafting committee. Eventually this led to a set of articles published in the *Hasting's Center Report*, which may be the first collection of essays on the ethics debate surrounding stem cells.[5] In the meantime, what the situation called for was new to these ethicists—they had to work in secret. Could there be such a thing as secret ethics?

The three-month secret period included additional shotgun tutorials to educate the ethicists on the science of stem cells. The board worked vigorously with the assumption that only good science could lead to good ethics. It would be necessary for the group to comprehend the science before it could have confidence in its formulation of the relevant ethical issues. This concern later inspired Karen Lebacqz to write an essay, "Bad Science, Good Ethics."[6]

Working under constraints of time, diversity, and secrecy, the EAB did its best to outline crucial ethical issues and to propose guidelines for stem cell research and offer justification for those guidelines.

As the EAB began its work, the first thing we tenderfoot ethicists needed was an education in the science of stem cells. Roger Pederson of the University of California medical school along with Geron's Tom Okarma, then director of research and at this writing CEO, became our teachers. It was a crash course.

Deriving Human Embryonic Stem Cells

We started with derivation, both of hES cells and primordial hEG cells. While we will be primarily concerned with hES cells here, we start with a few words about hEG cells.

John Gearhart and his colleagues at Johns Hopkins dissected abortuses between five and eight weeks old. They removed the contents of the primordial gamete ridge. Eggs here still have the full complement of 46 chromosomes, and they are undifferentiated. They are pluripotent. It is not known yet whether the cells derived from this technique have all the characteristics of hES cells. However, the technique may be important ethically, as the scientist deals with embryos that have already been "destroyed" by someone else; for some commentators, this makes the technique more ethically acceptable. Nonetheless, in light of the significance of hES cells, our EAB gave proportionately more attention to what was happening at the University of Wisconsin.

At Wisconsin, James Thomson was establishing four characterized stem cell lines. Three would be derived from frozen and thawed zygotes. One would begin with a freshly fertilized egg. Once activated, the zygote would divide into multiple totipotent cells. At the blastocyst stage between four to six days, an outer layer would appear. This outer layer is the trophectoderm, a sort of crust that could eventually produce a placenta to enable adhering to the uterine wall of the mother. The target was the inner cell mass, the as yet undifferentiated stem cells.

The trophectoderm would be broken, and the cells of the inner cell mass would be placed on a feeder tray. These early stem cells were telomerase enhanced. So, they became immortal in the sense previously described. As immortal, they are called "characterized."

The activated egg already possesses 46 chromosomes, the full human complement. When a woman's egg is fertilized by a man's sperm, each contributing 23 chromosomes, it becomes a zygote that includes genetic material from both the man and the woman (see figure 2.1). The primary source of zygotes for stem cell research is leftover IVF eggs frozen in fertility clinics.

The petri dish or feeder tray originally provided nutrients for the cultured stem cells in the form of ground-up mouse fibroblasts. Out of fear of contamination of human cells with materials from animal cells, the characterized lines were later removed from contact from mouse nutrients. The Wisconsin cell lines have been decontaminated, so to speak.

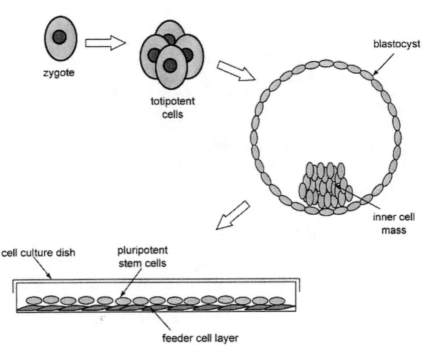

Figure 2.1. The derivation of hES cells. Diagram by Martinez Hewlett, used with permission.

As it turns out, Thomson won a race that was being run by scientists in Australia, Singapore, and Israel. Building on the work of Singapore's Bongso, Pacific and Israeli researchers were prospecting for the very gold that Thomson found. In fact, Thomson had actually borrowed research material from one of his Israeli competitors. Cynthia Fox waves the finish line flag with these remarks: "The tightness of this race, and the slimness of the margin with which Thomson won it, was breathtaking, especially in light of another irony: most of Thomson's cells were actually from Israel, cultivated with the help of the Haifa lab of Joseph Itskovitz-Eldor."[7]

Later in the fall of 1998, back in Boston at ACT, Michael West was also racing forward. West reported that he had caused human somatic cells to revert to the undifferentiated state—returning to pluripotency—by fusing them with cow eggs. This fusion produced cells resembling ES cells.[8] West's experiment was reported in the press, not a peer-reviewed journal. We at the Geron EAB were apprised of the West research, but we concentrated primarily on what had been accomplished at the University of Wisconsin and, to a lesser extent, Johns Hopkins.

The Ethics Scramble

The Geron EAB began with a scramble. The first two meetings in the summer of 1998 included input from Roger Pederson and Tom Okarma. We were told that stem cell research was being conducted at Geron in accord with guidelines set by two government reports: the 1994 report of the Human Embryo Research Panel (HERP) and the 1997 report and recommendations on cloning by the National Bioethics Advisory Commission (NBAC). A moratorium was in effect on any federal funding for research involving stem cells. Dolly the sheep had been cloned. The Society for Developmental Biology had voted for a voluntary moratorium on human cloning. This was the era when a doctor by the name of Richard Seed had announced that he would nonetheless pursue the goal of producing a human child through cloning. The cloning controversy swirled in the air.

In the midst of this social and political situation, the EAB turned to identifying some of the ethical issues that might be discussed in conjunction with Geron's stem cell program. During the summer of 1998 prior to the University of Wisconsin discovery, the Geron EAB prepared a list of ethical concerns, a list of areas for ethical research that could complement the scientific research. As the fall of 1998 rolled on, the EAB reexamined them in light of the fast-moving scientific frontier. Among the issues we identified at that time were the following:

- What would be the moral status of a blastocyst generated by somatic cell nuclear transfer rather than by fertilization? Would it be an embryo? Would it be a human with all the moral protections accorded human beings?
- Should human life forms and DNA be patentable?
- How should "great unintended consequences" or the possibility of unknown risks be factored into moral deliberation?
- What is the proper role of the media as science develops in new arenas?
- What are the risks in *not* pursuing scientific research? Who might suffer?
- What guidelines should govern this research, in the absence of a theory adequate to deal with new entities and procedures?
- What ethical issues surround the structures of research, the division of costs and profits?
- What justice and health concerns should we register on behalf of women who might consider donating eggs to scientific research?

What is significant in this list was that questions surrounding the embryo appear, but by no means claim exclusive ethical status. The question of the embryo appears as one item on a long list.

At the initial meeting we thought we would have plenty of time to ferret our way through this list, believing that procuring stem cells was still just an idea. Without alarm or urgency, as reported in the previous chapter, the EAB chose Karen Lebacqz as its first chair and set the next meeting for late September.

Before that September meeting could take place, however, the University of Wisconsin bombshell was dropped. We were summoned and told, in strictest confidence, about James Thomson's work under the auspices of Geron. No Eden of calm ethical deliberation would be enjoyed. We were thrown out of the garden of academic speculation into the wilderness of highly specific, pragmatic, and consequential tasks. It was our task to formulate as quickly as possible what we believed Geron should be addressing as stem cell research moved toward experimentation and therapy. Okarma said that in November of that year the Wisconsin and the Johns Hopkins discoveries would be announced. We would have two months to consider our ethical position on the derivation of hES and hEG cells and to establish guidelines for Geron research on these cells. We scrambled to get ready.

Because time was of the essence, we developed the following process:

- With the help of a research assistant,[9] we culled the existing literature[10] on the status of blastocysts or "pre-embryos," as the developing zygote was commonly called at that time.

- Four members of the Geron EAB wrote or provided from their previous work a "position paper" on some crucial background ethical issue that might be applied to stem cell research.[11]
- The chair of the EAB, Karen Lebacqz, then drafted a memo regarding issues that should be included in a position paper. These issues were:

1. The status of the blastocysts from which stem cells are derived.
2. Proprietary rights in cell lines—do donors of eggs or embryos have any proprietary rights in the cell lines that might be developed from donated tissue?
3. What kind of "informed consent" should be necessary for use of IVF embryos?
4. Ethical issues in the conduct of the research—should limits be set on what research can ethically be conducted; should there be special requirements for destruction of tissues or cell lines; and so on?
5. Ethical issues in the long-term consequences of the research, including costs and benefits.
6. The role of the EAB, especially regarding our interactions with media and our public accountability.

At the September meeting, we agreed to provide both a brief position paper setting out guidelines for ethical conduct of stem cell research and a background paper to provide the warrants for the ethical guidelines that we proposed. These guidelines and the supporting warrants were later published in the *Hastings Center Report*.

The Guidelines and Warrants

The EAB was a diverse group, including theologians from Jewish, Roman Catholic, Lutheran, and Calvinist backgrounds as well as one ethicist who identifies primarily as a secular thinker. Yet in spite of our diversity, we were unanimous in our belief that stem cell research could be conducted ethically. Indeed, we had little difficulty agreeing on some basic guidelines for stem cell research, so long as those guidelines were stated in somewhat general terms. We stated them as follows:

1. The blastocyst must be treated with the respect appropriate to early human embryonic tissue.
2. Women or couples donating blastocysts produced in the process of IVF must give full and informed consent for the use of the blastocysts in research and in the development of cell lines from that tissue.

3. The research will not involve any cloning for purposes of human reproduction, any transfer to a uterus, or any creation of chimeras.
4. Acquisition and development of the feeder layer necessary for the
   growth of hES cells *in vitro* must not violate accepted norms for human
   or animal research.[12]
5. All research must be done in a context of concern for global justice.
6. All such research should be approved by an independent EAB in addition to an Institutional Review Board.

These guidelines can easily be seen as reflecting three basic ethical principles that have long been accepted as foundational for bioethics—respect
for persons, beneficence, and justice. "Respect for persons" was a principle
identified by the National Commission for the Protection of Human Subjects of Biomedical and Behavioral Research in the now-famous *Belmont
Report*. This principle gives rise to requirements of informed consent, as
demonstrated by classic texts such as Tom Beauchamp and James Childress's
Principles of Biomedical Ethics[13] and Ruth Faden and Tom Beauchamp's *A History and Theory of Informed Consent*.[14] The notion of respect and its translation into requirements of consent is exemplified in guidelines 1 and 2 of the
EAB, which require both the informed consent of the woman or couple who
donate blastocysts, and respect for those blastocysts.

Beneficence—doing good—is the second foundational principle of bioethics. It is often partnered with the principle of non-maleficence, or not-
harming. Some ethicists separate these principles into two. Others conflate
them. In their best-selling textbook, Tom Beauchamp and James Childress
have typically addressed them separately, but Robert Veatch, in contrast,
discusses them as two sides of the Hippocratic oath.[15] Beneficence and non-
maleficence are seen in requirements for "cost-benefit" analysis, or for benefits sufficient to outweigh burdens.

The principle of beneficence and its accompanying concern for non-
maleficence is reflected in guidelines 3 and 4. Guideline 3 protects against
bringing to birth a child who would face unknown risks from cloning. Guideline 4 is a rather general statement regarding rules for research on humans or
animals and reflects general concerns to abide by principles of cost-benefit
or harm-benefit ratios.

Justice is stated clearly in guideline 5. However, as one critic pointed
out, the EAB did not elaborate carefully what justice might require in stem
cell research.[16] Recently, several other groups have taken up some of the
problems of justice involved in stem cell work.[17] One glaring problem is
money: stem cell research is big science, requiring big money. Only investors

gambling on a profit have shown a willingness to support this research. So, an ethicist must ask: What will this mean for access to the products? If the therapeutic products of stem cell research turn out to be expensive, how will those near the bottom of the economic ladder benefit? Will they have access to the medical benefits? Despite the significance of such questions, to date they remain unanswered.

As time has passed and the ethical discussion has expanded, the key word to indicate that justice is at stake is *access*. Access applies at the first stage of research and the final stage of distribution. At the first stage, patenting becomes the issue, and we will address this specifically in a later chapter. Does privatizing discoveries with Intellectual Property (IP) protection through patenting restrict the rapid spread of progressive research, or does it enable the sharing of research results, secure in the knowledge that others cannot "steal" one's invention? To make the sharing of research materials easier, some funders require that researchers who derive new stem cell lines immediately place their lines into the public domain. They do this by placing samples of new lines in stem cell banks, so that scientists all over the world can borrow such lines for their own research. In the future, when therapies have been developed, ethicists will want wide access to the benefits of the research, universal access if possible. Access at the beginning and the end is the way the justice concern is becoming formulated.

Finally, the Geron EAB thought the ethical issues involved in stem cell research were sufficiently new and distinctive that scientists should institute a practice of independent ethical review in addition to the usual requirements for Institutional Review Board (IRB) approval. This turned out to be prescient. When in 2005 the National Academies of Sciences (NAS) issued its guidelines for stem cell research, this became a recommendation. NAS recommended that each university set up an Embryonic Stem Cell Research Oversight (ESCRO) committee that would bring specialized expertise, including ethics, into the monitoring of research protocols.[18] In some institutions this has been shortened to Stem Cell Research Oversight (SCRO) so as to include an array of stem cells that do not strictly fit into the embryonic category. The key point is this: a primitive yet valuable linking of scientific and ethical arms has begun to take place.

"Embryo" versus "Blastocyst"

Although the EAB was unanimous in its belief that stem cell research can be conducted morally, we nonetheless thought that the ethical issues involved in stem cell research were new and troubling. One of the frustrations we faced from the beginning is that so much of the discussion of the status of developing human life would be taken from the abortion debate. The EAB was careful at the outset to use the term "blastocyst" when talking about the developmental stage at which stem cells are derived. Note, for example, that our first guideline above specifies respect for the blastocyst and the second guideline talks about donation of blastocysts. We discussed the moral status of the blastocyst, not of "embryos," as much of the contemporary discussion does.

Yet the cells resulting from the isolation and culturing of the inner cell mass had been identified in the scientific literature as ES cells. This term became the commonly accepted term. While we debated at some length whether the blastocyst or the resulting stem cells should be identified as "embryonic," ultimately—and perhaps unfortunately—we capitulated to common usage.[19] We were convinced immediately that the stem cell matter should be sharply distinguished from the three-decades-old abortion debate, but we did not know exactly how to word our concern. Despite our belief then that the discussion of stem cells *should* be separated from the abortion debate, alas, as time passed it became impossible to do so.

Guidelines and Rationales

While it was not difficult for us in the Geron EAB to agree on six guidelines, it was sometimes difficult to agree on how to express the warrants or justification for those guidelines. An initial draft of guidelines and supporting rationale was prepared by Karen Lebacqz and Michael Mendiola during the fall of 1998. The EAB met in December to discuss the draft. We had agreed to a March 1999 deadline in order for our guidelines to be published in the *Hastings Center Report*.[20] The December meeting was tense! Because we had agreed so easily on a preliminary set of guidelines, we wanted to agree as well on the statement of justification for those guidelines, rather than leaving each member free to write his or her own "warrant" statement. But reaching agreement on the background statement proved difficult.[21]

For instance, we say in the statement that we affirm moral status as developmental. This is true. But it is also true that some of us would assign "personhood" or protectable moral status at conception while others would not assign it until much later in development. What it meant to operate out of a developmental framework could differ significantly from member to member of the EAB. As a group we were able to affirm, however, that "respect" applies as an ethical principle at *every* stage of development. This left each member some freedom in understanding exactly what "respect" would mean.

We could agree on one core meaning—that in the absence of a capacity for sensation (pain) or thought, the blastocyst was respected by being sure that it is used only with care and in research having "substantial" value. Of course, what constitutes "substantial" value could also be open to dispute. We suspect that no member of the EAB was entirely satisfied with every statement in the final essay. Everyone would have said something differently. But at least we were able to contribute to the public discussion by displaying a set of guidelines and providing some defensible warrants for those guidelines.

Histocompatibility and Nuclear Transfer

Note that one of our guidelines proscribes reproductive cloning. Why would cloning become an issue?

It becomes an issue because of the histocompatibility problem—that is, the challenge of making patient-specific lines of stem cells. The goal of deriving stem cells, recall, is to tease these undifferentiated cells into becoming specific tissue. Once they have committed themselves to becoming a specific tissue, then the tissue could be surgically transplanted into a patient to regenerate an organ. Such a plan would be foiled, of course, if the recipient's body would reject this new tissue.

Each cell has glycoproteins on its surface membrane. The body's immune system can read the configuration of these surface proteins and recognize the difference between its own cells and foreign invaders. The immune system puts up a defense against the foreigners. How can surgeons slip new regenerative cells into a recipient's body and get past the immune defense?

If the stem cell tissue would be the same genetic code as the cells already in a person's body, then the immune system would not be triggered. The surface proteins would match. The invaders would be deemed friends. But, how could we create a stem cell line with a matching genome? Answer: cloning.

The key to Ian Wilmut's success with bringing Dolly the sheep into the world was *somatic cell nuclear transfer* (SCNT), or *nuclear transfer* (NT) for short. Wilmut returned some mature DNA to its undifferentiated state, then he placed it in an enucleated egg, an oocyte. He transferred the DNA nucleus of one sheep into the egg of another sheep. Once the egg with the transferred undifferented DNA was activated and placed in the uterus of a third sheep, it produced a lamb with the genetic code of the original DNA source.[22] Could this be done to make stem cells? Could we consider cloning for therapeutic purposes?

Tom Okarma at Geron thought this would be worth exploring. So he went to Scotland and negotiated with the Roslin Institute. Wilmut became our colleague, so to speak. So did Dolly. A large picture of Dolly was hung in the Geron board room, where the EAB could look at her while deliberating.

As the experiments proceeded, it appeared increasingly to Okarma that NT would not solve the problem adequately. Yes, therapeutic cloning could attain histocompatibility. But, would it be worth the cost? In 1995 Wilmut had failed 277 times before he hit success with Dolly. By the year 2001, his ratio of success was 100 to 1. Okarma was dejected by the slowness of the technological advance in NT techniques. He tried to envision all the patients around the world needing tissue transplants, each requiring 100 failures at NT before a patient-specific stem cell line could be established. A giant laboratory with expensive equipment and many employees would be required onsite for each patient. The expense and inefficiency would be intolerable as a generally available therapy. NT did not look like a scalable technique. Another road toward histocompatibility would have to be taken.

The map to histocompatibility at one point contained six alternative routes. Sparing you the details on each, they are: (1) somatic cell nuclear transfer, (2) cell line banks, (3) universal donor cells, (4) chimerism, (5) parthenogenesis, and (6) adult stem cells. The EAB took upon itself a thorough examination of all six. This study took two years of background research, writing, and editing. To the best of our ability, we drew on current bioethical resources and drafted what we believed would be a helpful document. But, when it became a near final draft it was summarily rejected by Tom Okarma. The science was moving so fast that our analysis was already out of date. Our failure on this task contributed to the decision to disband the EAB.

As of this writing, the most scalable and realistic route to histocompatibility is likely to be a combination of chimerism and cell banks. Preliminary findings show that stem cells injected into the bone marrow travel throughout the body, lodging in various organs. With this knowledge in mind, the theory goes like this. We could establish a bank of hES lines. The line with the genetic code most like that of the patient would be selected, and a sample withdrawn from the bank. Some of these stem cells would first be injected into the patient's bone marrow. Other stem cells *ex vivo* would be teased into specific tissue such as liver tissue. After two weeks, the invading stem cells will have traveled throughout the body, some lodging in the liver. When the surgeon inserts the new liver tissue, the already present stem cells in the liver will welcome them home.

We use the term "chimera" here because the patient would thereafter have two sets of genes, the original he or she was born with plus that of the stem cells now residing in the body. A *chimera* can be defined as an animal or person with two genotypes in stable coordination. This proposed procedure would result in a single person with two genotypes cooperating with one another. We think of this as *intra-species chimerism*, not a form of mixing species. No nonhuman animal genes would be introduced into a human system with this procurement method.

Implications for scalability should be obvious. The entire world could be supplied with a finite number of stem cell lines housed in a single bank. Physicians onsite could draw out of this bank a stem cell line most appropriate for a specific patient. Perhaps some small doses of immunosuppressants might have to accompany the therapy, but this may be tolerable. The cost would plummet, when compared to somatic cell nuclear transfer for each recipient. Scalability seems to be the scientific step we take toward economic justice.

Elsewhere, the route toward NT is still being followed by other researchers. Woo Suk Hwang and Shin Young Moon and their team at Seoul National University in Korea announced a significant breakthrough in 2004. They initially claimed to have produced 11 human stem cell lines that were genetic matches of 9 patients ranging in age from 2 to 56. Patient maladies included diabetes and spinal cord injuries. The success rate dropped dramatically to only 17 failures for each success, on average. If others would follow this model, then NT therapy could begin with an unfertilized egg rather than a zygote. Inserted into it would be the desired DNA. Then it would be activated. Cell division would take place. At the blastocyst stage, scientists would disaggregate the inner cell mass.[23] From this pool of cultured stem cells, cells for tissue cultivation would be drawn.

Such good news! However, the good news did not last long. It turns out that the Seoul National University claims were eventually proven fraudulent. So, as of this writing, success in nuclear transfer with human cells is still but a promise. Yet, the hope persists.

Is the promise on the near horizon? Possibly. In 2007 the first confirmable nuclear transfer in primates was announced. A team led by Shoukhrat Mitalipov of the Oregon Health and Science University took a skin cell from a nine-year-old rhesus macaque named Semos. They returned the skin-specific or differentiated DNA back to its predifferentiated state, preparing it for reprogramming. They placed the DNA nucleus into an enucleated oocyte. This led to the creation of two cell lines that were still holding their pluripotency many months later. "Our results represent successful nuclear reprogramming of adult somatic cells into pluripotent ES cells and demonstrate proof-of-concept for therapeutic cloning in primates."[24] With this achievement in primates, many scientists believe that nuclear transfer is just around the corner for humans as well.[25]

In January of 2008 a San Diego company, Stemagen, claimed to have produced embryo-like bodies by depositing the nucleus of a human skin cell into an enucleated human oocyte. Stemagen used 25 eggs to obtain 5 embryos which grew to clusters of between 40 and 72 cells.[26]

Even if nuclear transfer becomes successful in making human stem cell lines, there is still a question as to whether NT could become scalable. Michael West sides with NT advocates and believes that NT is still the most promising path to histocompatibility. Whether this will be proven remains to be seen. It is not the province of ethicists to bet on one scientific horse or the other.

Nevertheless, the EAB needed to address the cloning issue. We found ourselves in agreement with the worldwide scientific and religious consensus, namely, cloning for purposes of reproduction should be banned.[27] However, we argued that it would be necessary to approve cloning—understood as somatic cell nuclear transfer or NT—for laboratory scientists to proceed toward developing stem cell therapies. We still believe this.

Notes

1. Michael D. West, *The Immortal Cell: One Scientist's Quest to Solve the Mystery of Human Aging* (New York: Random House/Doubleday, 2003), 189.

2. James A. Thomson, et al., "Embryonic Stem Cell Lines Derived from Human Blastocysts," *Science* 282 (1998): 1145–47.

3. Ariff Bongso, et al., "Isolation and Culture of Inner Cell Mass Cells from Human Blastocysts," *Human Reproduction* 9.11 (1994): 2110–17. Knowledge of the existence of stem cells in mice was first announced in A. J. Becker, et al., "Cytological Demonstration of the Clonal Nature of Spleen Colonies Derived from Transplanted Mouse Marrow Cells," *Nature* 197 (1963): 452–54.

4. Michael J. Shamblott, et al., "Derivation of Pluripotent Stem Cells from Cultured Human Primordial Germ Cells," *Proceedings of the National Academy of Sciences* 95 (1998): 13726–31.

5. See the "Symposium: Human Primordial Stem Cells," which includes the Geron Ethics Advisory Board analysis and commentaries by Glenn McGee and Arthur L. Caplan, Lori P. Knowles, Gladys B. White, Carol A. Tauer, and Lisa Sowle Cahill. *Hastings Center Report* 29.1 (March–April 1999): 30–48.

6. Karen Lebacqz, "Bad Science, Good Ethics," *Theology and Science* 1.2 (2003): 193–201.

7. Cynthia Fox, *Cell of Cells: The Global Race to Capture and Control the Stem Cell* (New York: W. W. Norton, 2007), 81.

8. Nicholas Wade, "Researchers Claim Embryonic Cell Mix of Human and Cow," *New York Times*, November 12, 1998, A1, A26.

9. Rachel Metheny, a Ph.D. candidate in the Graduate Theological Union, served as our research assistant for several months. Later, Gaymon Bennett, also a student in the Graduate Theological Union, became the research assistant to the EAB and attended almost all discussions.

10. Among the essays we reviewed were Thomas A. Shannon and Allan B. Walter, "Reflections on the Moral Status of the Pre-Embryo," *Theological Studies* 51 (1990): 603–26; Thomas A. Shannon, "Cloning, Uniqueness and Individuality," *Louvain Studies* 19 (1994): 283–306; Carol A. Tauer, "Personhood and Human Embryos and Fetuses," *Journal of Medicine and Philosophy* 10 (1985): 253–66; Lisa Sowle Cahill, "The Embryo and the Fetus: New Moral Contexts," *Theological Studies* 54 (1993): 124–42.

11. The papers were Michael M. Mendiola, "Some Background Thoughts on the Concept of 'Moral Status' Relative to the Early Embryo"; Laurie Zoloth-Dorfman, "The Ethics of the Eighth Day: Jewish Bioethics and Genetic Medicine"; Ernle W. D. Young, "Draft Statement on the Moral Status of the Human Embryonic Tissue"; and Ted Peters, "When Does an Individual Human Life Begin?"

12. At the time when we first developed these guidelines, the "feeder layer" consisted of irradiated mouse embryo cells; hence, the concern for ethical research involving animals. However, the cell lines no longer use that culture medium and have indeed been free of any animal cells for many generations.

13. Tom L. Beauchamp and James F. Childress, *Principles of Biomedical Ethics*, 4th ed. (New York: Oxford University Press, 1994).

14. Ruth R. Faden and Tom L. Beauchamp, *A History and Theory of Informed Consent* (New York: Oxford University Press, 1986).

15. Robert M. Veatch, *A Theory of Medical Ethics* (New York: Basic Books, Inc., 1981).

16. Lisa Sowle Cahill, "The New Biotech World Order," *Hastings Center Report* 29.2 (1999): 45–48.

17. Ruth R. Faden, "Public Stem Cell Banks: Considerations of Justice in Stem Cell Research and Therapy," *Hastings Center Report* 33.6 (2003): 13–27.

18. National Research Council and Institute of Medicine of the National Academies, *Guidelines for Human Embryonic Stem Cell Research* (Washington, DC: National Academies Press, 2005).

19. For example, Laurie Zoloth and Karen Lebacqz joined with Suzanne Holland in publishing a volume on the ethics of stem cells and entitled it *The Human Embryonic Stem Cell Debate* (Cambridge, MA: MIT Press, 2001).

20. Geron Ethics Advisory Board, "Research with Human Embryonic Stem Cells: Ethical Considerations," *Hastings Center Report* 29.2 (1999): 31–36; reprinted in *The Stem Cell Controversy: Debating the Issues*, edited by Michael Ruse and Christopher A. Pynes (Amherst, NY: Prometheus Books, 2nd ed., 2006) 117–29.

21. This should not really come as a surprise. In their wonderful study of casuistic—or case-based—reasoning, Jonsen and Toulmin note that it is often easier to agree on practical guidelines than on the reasoning behind them. See Albert R.

Jonsen and Stephen Toulmin, *The Abuse of Casuistry: A History of Moral Reasoning* (Berkeley: University of California Press, 1988).

22. I. Wilmut, et al., "Viable Offspring Derived from Fetal and Adult Mammalian Cells," *Nature* 385 (1997): 810–13.

23. W. S. Hwang, et al., "Evidence for a Pluripotent Human Embryonic Stem Cell Line Derived from a Cloned Blastocyst," *Science* 303 (2004): 1669–74.

24. J. A. Byrne, et al., "Producing Primate Embryonic Stem Cells by Somatic Cell Nuclear Transfer," *Nature* 450 (2007): 497–502.

25. David Cyranoski, "Race to Mimic Human Embryonic Stem Cells," *Nature* 450 (2007): 462–63.

26. "First Human Somatic Cell Nuclear Transfer Reported," www.scienceprogress .org/2008/01/first-human-somatic-cell-nuclear-transfer-reported/, accessed February 2, 2008.

27. In subsequent work, however, Lebacqz has given modest support for reproductive cloning under some circumstances.

~

CHAPTER THREE

Working within the Research Standards Framework

The beginning of our ethical deliberations at Geron's invitation was like putting together a large puzzle. We did not know how many pieces there would be. Nor did we know what the final picture would look like. Actually, it was more like cutting the puzzle pieces to shape before knowing just how they might eventually fit together.

Later, as the ethical picture began to come into focus, we could see better where things might belong. One category of ethical concern on our minds then we now identify as the *research standards framework*. This framework for defining and analyzing ethical matters has nothing overtly to do with religion or theology. Yet, it is a framework within which professional scientists and ethicists are likely to find themselves in communication with one another. Later in this book, we hope to show how research standards, though formulated in scientific and professional language, covertly reflect religious sensibilities and articulate secularized versions of theological principles.

Here in this chapter we will recall some of our initial forecasts about ethical issues we thought might arise. When they arose, intricate ethical inquiry emerged and dramatic events began to happen.

Egg-Donating Women

One of our guidelines emphasizes "informed consent" for women or couples involved with egg donation for purposes of scientific research. We also discussed the question of compensation. Should a woman be paid for the eggs she donates? As time has passed, this has become an important issue on the agendas of a number of ethicists.

What is it that this kind of research needs? What is the gold standard for egg donation? First, the highest quality eggs come from younger women. Women between 21 and 35 move to the top of the list. Mothers of at least one biological child increase the promise of quality eggs. Second, eggs should be free from infectious viruses. Screening out excess genetic material becomes part of the selection process. Third, experience shows that clinics like to work with women willing to participate, who genuinely want to share what their bodies produce for the wider human welfare.

What does this mean for women considering donating eggs for stem cell research? Some young women are selling their eggs to earn money to support them in their college educations. It appears to be lucrative. Yet, we note also that in recent years many women have offered themselves to clinics, willing to donate because they deem such research to be important. They voluntarily donate their eggs out of high-minded service. Making money is not the primary motive in many cases. Making a contribution to cure Parkinson's or cancer is meaningful. At least this is what donating women to date have reported. We might think of them as unsung heroines.

However, retrieving unfertilized eggs from a woman's body is complicated and risky. Hormonal manipulation is necessary to stimulate production of as many eggs as possible. Normally, when nature governs a woman's cycles, eggs appear and disappear in sequence. When clinically stimulated, however, the woman hyperovulates and produces a large number of eggs all at once. A dozen eggs can be retrieved at the same time. This makes for efficiency, to be sure.

Yet, there are health risks. Egg-donating women who later become pregnant might undergo ovarian hyperovulation syndrome. Such women could experience considerable pain. Remaining eggs may become enlarged, sometimes by a factor of ten. Bleeding can occur in the short term. There is also the risk of infection, which can cause sterility. And the risk of ovarian cancer may go up in the long term. Any woman contemplating egg donation will need to assess all of these possibilities. Respect for the woman's autonomy—*voluntary and fully informed consent*, among other things—is ethically required of the clinic and the researchers.

One of the big questions the Geron EAB posed, and which ethicists continue to ask, is this: Should women be compensated for egg donation? Our EAB voted "no." In 2005 the National Academy of Sciences similarly said "no," as did California's Proposition 71. What is meant here is that women donating eggs should not receive a fee, a profit. The core philosophical issue is that a woman's gametes seem to belong to what makes us human; and, due to the principle of dignity, what makes us distinctively human cannot be bought or sold. An associated issue is the fear of exploitation of women donors. The fear is that women might be tempted to risk their health because of financial rewards. Justice requires a particular concern for poor women, therefore.

The justice concern for poor women has been raised in the General Assembly of the United Nations. In 2003 and 2004 the question of banning either reproductive cloning or therapeutic cloning or both was raised. Representatives of many developing nations expressed fear that commercial use of cloning would create a demand for the eggs that poverty-stricken women might wish to sell. Nigerian envoy Felix E. Awanbor said he supported a total ban on both reproductive and therapeutic cloning because women from poorer parts of the world, "particularly Nigeria, are most likely to be at risk as easy targets to source the billions of embryos required for scientific experimentation."[1] When money becomes an incentive, sound judgment regarding human health can be put at risk.

For that reason, ethicists do not want money to become an incentive. Nor do ethicists want to encourage trafficking in human bodily material for payment. At the same time, it can also be exploitive to insist that women give their eggs without compensation, while men are routinely paid for sperm donation. The justice question regarding payment for human eggs is a complex one that will continue to be debated. Although we sided with those who argue that women should not be paid for their eggs, per se, we do believe that donating women should receive reimbursement for relevant expenses, travel, missed wages, and even child care. And, most importantly, they deserve thanks.

The Korean Scandal

In 1998 we had anticipated and sought to avoid what later happened in 2005. We sought to plug up ethical holes before they could sink the scientific ship.

The controversy over Woo Suk Hwang of Seoul National University in Korea proved that an ethical hole could sink a scientific boat. Dr. Hwang made worldwide headlines in 2004 and 2005 for successfully employing nuclear transfer to establish genetically specific stem cell lines for a number of patients. He refined the technique of nuclear transfer well beyond the achievements of Ian Wilmut at the Roslin institute, where Dolly the sheep was the single success in 277 tries; Hwang could log one success for each

17 tries. His team also cloned a dog named Snuppy, an Afghan who made the cover of *Time* magazine. Hwang became a national high tech hero. Korean Air declared him Korea's "National Treasure." He received a gush of government funding—the rough equivalent of US$132 million—and established the World Stem Cell Hub. His scientific ship was sailing at full speed ahead.

By late 2005, however, an ethical hole made the ship take on weighty water. Where did Dr. Hwang obtain his oocytes? Did he violate any ethical principles? Did he bribe or coerce them from his student assistants, or did he pay women to donate them? Hwang publicly stated in May 2005 that all eggs had been harvested from volunteers without payment. By November, this was disputed. An American collaborator, Gerald Schatten of the University of Pittsburgh, pulled out of his relationship with Hwang's new center, alleging breaches in ethical conduct. The director of the MizMedi Hospital in Seoul, from which Hwang's lab obtained some of its oocytes, disclosed that each of the women had been paid the equivalent of US$1,400. South Korea's Health Ministry also disclosed that two junior scientists had given their own eggs for research. The latter was seen as particularly egregious. Critics describe a university laboratory as hierarchical; in a society as influenced by the Confucian tradition as is South Korea, subordinates find themselves in a dependent relationship to their superiors. This dependency makes the situation *de facto* coercive. Had the Hwang research team exploited women to obtain their oocytes?

In addition, it appeared to critics that Dr. Hwang had been lying. In his own defense, Hwang suggested he did not know the source of the eggs and that later ethical standards were not in place when these eggs were harvested in 2003 and 2004. "Being too focused on scientific development, I may not have seen all the ethical issues related to my research," he said in a news conference on November 25, 2005.[2] Some researchers, wondering if the science was solid, proposed confirmation studies.

Eventually, the 11 patient-specific stem cell lines were disqualified because the evidence had been fabricated. No successful human nuclear transfer had been accomplished, even if the dog "Snuppy" turned out to be a legitimate clone. Gerald Schatten—who had previously suggested that human cloning might never be possible because of nuclear spindling problems, and who had helped Hwang rewrite his article for publication in *Science*—subsequently dissociated himself from the Hwang enterprise. A second American whose name appears among Hwang's collaborators, Jose Cibelli, also withdrew support for Hwang. Both Schatten and Cibelli have been exonerated from any complicity in falsification of evidence.

The sinking toward disgrace was met by attempts to bail Hwang out and to rescue the hero before he hit bottom. Choi Hee-Joo of the South Korean Health Ministry applauded the women donors who "voluntarily" offered their eggs "for the success of the research by sacrificing themselves." He went on, "The donations were made according to values consistent with Eastern culture, and shouldn't be looked at from the standpoint of Western culture." He announced that 33 high school girls have stepped forward to offer their eggs for the good of science.

"Hwang gets a public relations extreme makeover," wrote Jennifer Lahl in a blog for the Center for Bioethics and Culture Network. "I have never witnessed such a fast public relations blitz. . . . Interesting, in Hwang's defense to be restored to hero is the statement that we can't judge him by our western cultural ethics . . . in a country with a long history of abuses toward women."[3] Hwang's Korean culture could not protect him for long, however.

On May 12, 2006, Woo Suk Hwang was indicted on charges of fraud, embezzlement, and violations of bioethics law. Korean prosecutors claim Hwang misappropriated $2.99 million in government research funds and private donations. Evidently, Hwang withdrew large amounts of cash, placed the cash in bags, and then deposited the cash in other banks in an attempt to avoid a paper trail. By the time he was caught, Hwang had 63 accounts under various names. To cover up the embezzlement, he prepared false tax statements indicating purchase of cows and pigs for research purposes.[4]

Others on Hwang's research team were also indicted: Sung Keun Kang for fraud in procuring government grants; Sun Jong Kim for destroying evidence and obstructing research work; Byeong Chun Lee for fraud in procuring government grants and misappropriating funds; Hyun Soo Yoon for falsifying receipts and embezzling research funds; and Sang Sik Chang for bioethics law violations when procuring eggs.

The still popular hero, Hwang, has not lost all of his supporters. Hundreds of his disciples have mobbed the front of the prosecutor's office. On May 12, 2006, several Buddhist monks showed support for their hero by engaging in a 24 relay bowing ritual next to Jogye Temple in central Seoul. The Venerable Seoul, a Buddhist monk, stepped forward with a claimed $65 million to help Hwang restart his stem cell research. Cynthia Fox comments that the "unprecedented magnitude of 'Hwang-gate' will eventually be viewed by some as yet more evidence of the unusual allure of stem cells, of the unprecedented magnificence of their potential."[5]

Science is social and global. No longer can a researcher flee to the privacy of a laboratory. His or her work is tested not only in other laboratories but also in the public square. And the public demands that moral standards be upheld. Formulating appropriate moral standards that enhance the advance of salutary research while protecting human dignity and human welfare is the ongoing task of the ethicist.

Research Standards Needed

In what follows in this book, we explicate the three competing ethical frameworks for debating the morality of stem cell research: embryo protection, human protection, and future wholeness. What we see above is the need for standards to deal with investigative research and medical practices. We need a fourth framework—a research standards framework—to guide if not govern what scientists do in the laboratory.

There is scurrying the world over to establish and implement enforceable guidelines that deal with many of the above-mentioned concerns as well as a number of others. The government of Singapore published its guidelines in 2002; the *Japanese Guidelines for Derivation and Utilization of Human Embryonic Stem Cells* appeared in 2007. In the United States, the National Academies of Sciences issued a paradigmatic set of guidelines in 2005; the California Institute for Regenerative Medicine followed suit in 2006. As we proceed we will from time to time make reference to what is happening with the framework of professional ethics, what we might call the research standards framework.

However, our focus in this book is on those frameworks that purposefully ground themselves in a religious vision. The research standards framework presupposes a secular or nonreligious standpoint. What we will see, interestingly, is that in displaced fashion many religious concerns surface in secular language without theological footnotes. Even in laboratory legalese, the remnants of influential religious positions are visible. Like an exotic spice in a cream sauce, the taste of theology can still be discerned by the delicate scientific palate.

Conclusion

Back at Geron, despite the scramble at the beginning, our small team of ethicists zeroed in on the rapidly moving frontier of science and sought to provide insights into anticipated social consequences and ethical implications. We greeted positively the sense of public protection of women from exploitation.

Even though we viewed the question of the moral status of the blastocyst as important, by no means could we consider it the sole issue or even the most urgent. As stem cell research continues today, we regret and find frustrating how the public discussion seems to focus almost exclusively on questions surrounding the protection of the early embryo and obscures other vital matters such as the enormous potential for improving human health and well-being that this science might eventually deliver. To watch groups organize to prevent the forward movement of medical science with its potential for relief of human suffering is, for us, a significant ethical concern.

Notes

1. Cynthia Fox, *Cell of Cells: The Global Race to Capture and Control the Stem Cell* (New York: W. W. Norton, 2007), 81.

2. James Brooke with Choe Sang-Hun and Nicholas Wade, "Korean Leaves Cloning Center in Ethics Furor," *New York Times*, November 30, 2005, www.nytimes.com/2005/11/25/international/asia/25clone.html?emc=etal&pagewante.

3. Jennifer Lahl, "Hwang Gets a Public Relation Extreme Makeover," www.cbc-network.org/enewsletter/index_11_30_05.htm.

4. D. Yvette Wohn and Dennis Normile, "Prosecutors Allege Elaborate Deception and Missing Funds," *Science* 312 (2006): 980–81.

5. Fox, *Cell of Cells*, 12.

~

CHAPTER FOUR

Three Contending Frameworks
for Stem Cell Ethics

It was the Saturday former president Ronald Reagan died: June 5, 2004. We noticed something while attending a conference at the University of California, a conference Ted Peters had helped behind the scenes in planning. Gaymon Bennett was a featured presenter. Karen Lebacqz attended. The conference was hosted by SCAN, the Stem Cell Action Network. Of the nearly 300 people seated at tables, 100 were in wheelchairs or on crutches. When they spoke, they spoke passionately of their suffering relatives and of their own genetically based maladies such as multiple sclerosis, Parkinson's, diabetes, cancer, heart disease, and, yes, of course, Alzheimer's, the one that caused the deterioration and death of Mr. Reagan. As the latest research on stem cells was being presented, we could feel a sense of hope buoying group. "Science as savior," Lebacqz and Peters commented to one another.

Just the day before, June 4, 58 U.S. senators (43 Democrats and 15 Republicans) signed a letter to President George W. Bush asking the chief executive to relax federal restrictions on laboratory research. During that week Nancy Reagan appeared at a fund-raising dinner in Los Angeles to promote research on stem cells. On June 5 Senator Diane Feinstein of California told the press, "This issue is especially poignant given President Reagan's passing. Embryonic stem cell research might hold the key to a cure for Alzheimer's and other terrible diseases." In a press release a week later presidential hopeful John Kerry joined the chorus: "The medical discoveries that come from stem cells are crucial next steps in humanity's uphill climb."

To quell what appeared to be a rising tide of support for stem cell research, Wesley J. Smith of the Discovery Institute and the Center for Bioethics and Culture wrote a widely distributed e-letter, saying, "The intensity of belief in science as savior, combined with a desperate desire that it be so, has become so fervent that faith in this research has come to resemble a secular religion. And now, supporters of cloning for biomedical research are using the death of Ronald Reagan from complications of Alzheimer's disease as a bellow to blow the political winds in their favor."[1] Smith's was not the only voice being raised against laboratory research on embryonic stem cells. A cacophony of Vatican ethicists, American evangelicals, and secular naturalists have spoken out, and what they say has been translated into national executive policy.

At the SCAN conference,[2] two of the nation's leading researchers—one a preeminent expert on adult stem cells, Irving Weismann, and the other an expert on embryonic stem cells, Thomas Okarma—described experiments that displayed the hitherto inconceivable promise of tissue and organ regeneration. While describing the research, Weismann could hardly resist taking swipes at the White House for listening to those recalcitrant "religious" voices who want to halt the forward movement of science. When it came time for the theologian to speak, one of this book's coauthors, Gaymon Bennett, probably should have been wearing a dartboard target. Partially out of self-defense, Bennett adroitly mentioned to the audience that more than one ethical position on stem cell research could properly be called "religious." And he, as a Christian, could ethically support stem cell research because of its potential for relieving human suffering and enhancing human health and well-being. His message: There is more than one way to be moral, more than one way to translate one's faith commitments into public policy.[3]

Public Theology and the Need for Ethical Coherence

We find fascinating that, today, theology has again become a topic of public discourse. It is no longer limited to Sunday school. Theology has become a subject of discussion in scientific laboratories and in the halls of state. Some antireligious humanists and neosectarians want to limit theological discourse to the church. They feel that moral deliberation is supposed to be a private matter, a matter solely for vested religious interests. For these, renewed public interest in theology comes as bad news. But the fact is that we are all discussing theological perspectives. Even atheists find themselves in the discussions. For good or ill, theology is a topic in the public sphere. Helpfully or unhelpfully, religious positions justify ethical positions. These positions are influencing research protocols, national policy, and the future of medical care for millions of people.[4]

Given these developments we must ask: How should theology inform ethics? A plurality of ethical positions is fueling the fires of debate over morality in science and morality in government. Christians, Jews, Muslims, and others must carefully consider the relationship of fundamental faith commitments to the ethics of biomedical research. This is a time for religious leaders to recognize their responsibility to deliberate thoroughly and speak carefully and publicly. It is time for a reexamination of foundational ethical commitments.

Such reexamination leads to what follows in this chapter. While ethicists, theologians, politicians, and others fight in the public square over stem cell research, we want to step out of the fray for a few moments of reflection and deliberation. What should be our basic point of departure for ethical analysis and construction?

Part of the challenge here is that existing moral arguments in the public sphere work out of differing conceptual and ethical frameworks.[5] What is more, within each framework multiple positions can be taken up and defended. Now, just what is an ethical framework? An ethical framework is a way of selecting and arranging the salient elements of a social situation in such a way as to justify a moral position. The framework limits what is admitted for consideration. What is admitted is then organized around a focal ethical issue. The focal issue combined with a limit on the considerations admitted makes it easier to establish coherence in moral justification.

An ethical framework can be distinguished from a given position within that framework. Within any single framework we often find two or more moral positions. When disagreements arise within a shared framework, the disagreeing parties share enough basic presuppositions to understand and challenge one another. Yet, when the public discussion includes multiple frameworks, we talk past one another without actually understanding one another. The acrimony and demonizing that we hear are in large part tied to a frustration deriving from this confusion over frameworks.

We often find ourselves at an impasse. As Boston College ethicist Lisa Sowle Cahill puts it, "Public debate sometimes seems to be caught in an impasse between the value of embryos and the promised benefits of stem cell research"[6] Cahill rightly recognizes the impasse. However, she wrongly attributes it to a conflict of values. We, in contrast, believe the values may very well be compatible; yet they come to expression in competing ethical frameworks. When we fail to see the incompatible frameworks, what appears to be a conflict of values hides the existence of incompatible ethical frameworks.

The presence of incompatible ethical frameworks leads us to pause and ask: How is ethical deliberation actually being conducted today? Where are the points of convergence and divergence between frameworks? How should we sort through positions so that commitments and grounds can be examined? Our goal here is not to develop a comprehensive framework that would unite all the others; rather, we wish to supply an analysis to achieve a higher level of clarity. Once we gain some clarity regarding the existing impasse, we ask the constructive questions: Is it possible for Christians and other people of faith to recognize and even reconcile basic differences? Could we as a worldwide community come to share ethical commitments regarding the future of regenerative medicine? In proceeding toward these goals, we will work with a combination of hope and realism, somewhat in the spirit of Lutheran ethicist Robert Benne, who writes: "The church should provide a vision of the common good that is both hopeful and realistic."[7]

Three Ethical Frameworks

In what follows, we will analyze the three competing predominant frameworks within which the current public controversy over stem cell research is being argued: (1) the *embryo protection framework*, (2) the *human protection framework*, and (3) the *future wholeness framework*. Within each framework we can identify more than one coherent position. Just as a picture frame puts a border around the picture without determining the content of the picture, so also do these ethical frameworks set the parameters of moral arguments without determining the conclusions. We will show how this works in the stem cell debate as well as provide a balanced presentation of ideas with appropriate critique. We do not see all of the frameworks and positions as equally legitimate; yet, we want to present them fairly. After comparing these three frameworks, we will turn to developing support for our own position, saying "yes" to stem cells, which we argue for from within the future wholeness frame. However, we note that it is possible to support stem cell research from within any of these frameworks, contrary to popular opinion. Before advancing our own position, we will provide examples of how moral arguments can talk past one another if they do not recognize the alternative ethical frameworks.

We have been refining our vocabulary as we have analyzed the global debate. The second framework (the human protection framework) was previously called the "nature protection framework." This is because the arguments within this framework are naturalistic in character—that is, they appeal to our inherited nature for their source of value. However, upon a closer look, we can see that the emphasis is clearly on *human* nature, not nature in general. Human nature protectionists are concerned about defending us from dehumanization by the science of genetics. Hence, the new label, the "human protection framework."

Similarly, we now use the term *future wholeness framework* to replace what we previously called the "medical benefits framework." Arguments on behalf of medical benefits are a subcategory within the larger vision of a new future for human well-being. Using the language of "benefits" may sound utilitarian, but those who use this framework are not necessarily utilitarian. The central commitment within this framework is to a vision of a future that will differ from the past. Scientific research can lead to relief of human suffering. Yet, the theologians would add, as important as medical benefits might be, human flourishing includes much more than physical health alone. It includes spiritual wholeness as well as physical wholeness. These two belong together, and we need a label that clearly incorporates both.

So, in sum, we distinguish three principal frameworks in the stem cell debate. Two frameworks emphasize protection—the protection of the embryo and the protection of our natural humanity. The third framework orients us toward future transformation (see table 4.1).

In addition to these three, professional scientists work within a fourth framework, which we label the *research standards framework*. Unlike the three principal frameworks, which are often informed by theological commitments, this fourth is an avowedly secular or nonreligious framework. However, reading between the lines will reveal some religious subtexts; theological flavor can yet be tasted in the scientific fare. This is not a problem in itself. Our task is simply to identify how and where theological reasoning informs the discussion so as to think more clearly and carefully about the stakes of stem cell research. Our treatment will deal directly with the first three frameworks and only indirectly with that of research standards.

Table 4.1. Three Theologically Based Ethical Frameworks

Embryo Protection	Human Protection	Future Wholeness
• "Pro-Life" on Abortion • Official Roman Catholicism • Evangelical Protestants • George W. Bush White House	• President's Council on Bioethics • Leon Kass • Anti-Playing God • Anti-Brave New World	• Medical Benefits • Most Research Scientists • Christian Authors of this Book • Jews and Muslims • SCAN

Ethical Storm Clouds on the Horizon

As we noted in an earlier chapter, Linda Sonntag had already seen ethical storm clouds on the horizon in 1996. This is why she refused to support stem cell research without sufficient ethical deliberation.

Michael D. West and his colleagues read the ethical weather report and were still ready to go out into the rain. When negotiating with the University of Wisconsin to pursue human embryonic stem cells, James Thomson reports saying to West at Geron, "I have visions of people protesting outside of your facility, with T-shirts saying, 'Stop the Killing of Little Beating Hearts.'" West heard Thomson's warning. "But," he said, "I so desperately wanted people to see what I saw—a power that could heal a broken heart, that could give children a cure for diabetes, that could regenerate cells and tissues that cannot normally heal themselves, like the nerves in the spinal cord and brain."[8]

What we see West doing here is shifting from the first to the third moral framework. Whereas Thomson posed the ethical problem as one of the moral status of the preimplantation embryo, "the killing of little beating hearts," West reframed the issue in terms of potential medical benefits to future patients.

Had West addressed Thomson directly at this moment, he could have remained within the embryo protection framework. He could have justified pursuing hES cells on the grounds that he does not deem the destruction of the blastocyst—the embryo at four to six days—as immoral. This is what West thinks. West is aware of fetal wastage in nature. He notes how 60 percent of zygotes or naturally fertilized eggs in a mother's body are expelled prior to implantation. The body discards the vast majority of them, for unknown reasons of nature. Having noticed how nature treats the early zygote, West patterns his moral judgment accordingly.[9] "Ethicists who study embryology are very comfortable with the idea of preimplantation embryos being either implanted or discarded in the way sperm is discarded if a pregnancy is not desired."[10] Within the embryo protection framework, we might refer to Michael D. West's position as the *morally permissible research position*.

Recall that Thomson at the University of Wisconsin is credited with the first isolation and characterization of hES (embryonic) cells, and John Gearhart is credited with the first isolation of hEG (germ) cells. When West was negotiating with Gearhart at Johns Hopkins, Gearhart anxiously foresaw "the collision course we were on with the religious right." West was less fearful. "I, on the other hand, because of my years of study in fundamentalist Christianity, could see the motivating of people on both sides of the issue.

Even so, John and I saw eye-to-eye on where this was all going. This research was going to unleash one of the fiercest battles between religion and science in recent history."[11] In this instance, West waffles. On the one hand, he sides sympathetically with the religious right, understanding fundamentalist theology and its motivations. On the other hand, he sees himself as a scientist at war with religion. He arms himself for battle. The mistake is that he sees this as a battle between science and religion, not as a battle between ethical positions on the role that science should play.

In the next few chapters we will describe in detail the three frameworks, showing how competing moral judgments and ethical commitments encourage and inhibit laboratory research on hES cells. Religious ethicists who judge hES cell research to be immoral are not necessarily antiscience, even though some voices within the human protection framework may sound like they have an antiscientific tone. Overall, ethical thinking grounded in a theological vision typically celebrates the advance of medical science. At the same time, such reflection attempts to guide scientific research in directions that will edify and not undermine human dignity and well-being.

Notes

1. Wesley J. Smith, "Cell Wars: The Reagans' Suffering and Hyped Promises," *National Review Online*, June 8, 2004.

2. For SCAN, see www.stemcellaction.org. For the "People of Faith for Stem Cell Research," see www.pfaith.org.

3. "We all read the same science and the same scriptures," writes Ronald Cole-Turner, "but we put them together with different formulas and different claims of authority for each. We all believe human life and human action are to be offered in obedience and praise to the same God, but we have profoundly different understandings of what God is doing in our time and requiring us to do." Ronald Cole-Turner "Introduction," in *God and the Embryo: Religious Voices on Stem Cells and Cloning*, ed. Brent Waters and Ronald Cole-Turner (Washington, DC: Georgetown University Press, 2003), 17.

4. "Religious convictions and apprehensions bear down on this tiny dot, the embryo, as a focal point for profound concerns about human identity, dignity, and manipulability. These intensely religious concerns mean the embryo is politically charged, and any analysis of public policy must take religion into account." Ronald Cole-Turner, "Introduction," *God and the Embryo*, 13–14.

5. We use the term "framework" where University of California cognitive scientist and rhetorician George Lakoff uses "frame." He writes, "Deep frames structure your moral system or your worldview. Surface frames have a much smaller scope. They

are associated with particular words or phrases, and with modes of communication." Important for us in analyzing the public controversy over stem cells is that frames structure what we take to be common sense; and this implies that "frames trump facts" in public debate. George Lakoff, *Whose Freedom: The Battle over America's Most Important Idea* (New York: Farrar, Straus and Giroux, 2006), 12–13.

6. Lisa Sowle Cahill, book review of *The Human Embryonic Stem Cell Debate: Science, Ethics, and Public Policy*, ed. by Suzanne Holland, Karen Lebacqz, and Laurie Zoloth (London: MIT Press, 2001) in *National Catholic Bioethics Quarterly* 2.3 (Autumn 2002): 562.

7. Robert Benne, *The Paradoxical Vision: A Public Theology for the Twenty-first Century* (Minneapolis, MN: Fortress Press, 1995), 222.

8. Michael D. West, *The Immortal Cell: One Scientist's Quest to Solve the Mystery of Human Aging* (New York: Random House/Doubleday, 2003), 152.

9. The pattern of "excess" in nature was also noted by Ernle W. D. Young on the Geron EAB, and was very influential in his support for stem cell research.

10. West, *Immortal Cell*, 146.

11. West, *Immortal Cell*, 157.

CHAPTER FIVE

The Embryo Protection Framework

"Choose a life affirming alternative," said U.S. President George W. Bush at a White House press conference on May 23, 2005. As an alternative to embryonic stem cell research, discarded *in vitro* fertilization (IVF) embryos should be adopted by families, implanted, brought to term, and become children. The president complimented the Nightline Christian Adoption program for providing the 21 children flanking him during the news conference.

To listen superficially to the public debate, one might get the impression that only one simple issue is at stake, the abortion issue in thinly disguised form. "Much of the theological debate about stem-cell research centers on the question of when life begins," says a sedate article in *Nature* magazine.[1] Pundit Ann Coulter thinks she knows when life begins: at conception. She bluntly accuses human embryonic stem cell researchers of being baby killers. "Embryonic stem-cell researchers are virtually never doctors. They're biologists. They don't care about healing people, they just want to be paid to push petri dishes around the lab, cut up a living human embryo, and sell it for parts like a stolen Toyota at a chop shop."[2]

The fire of the previous abortion debate continues to burn in the stem cell debate. Yet, a seismic shift in focus has taken place. During the 1970s, 1980s, and 1990s pro-lifers focused on the morality of removing a living fetus from the body of a mother. In the stem cell debate, no mother is present. The moral argument is over the blastocyst in a laboratory petri dish. Embryo protectionists contend that the blastocyst is a human person either actually or potentially and, as such, should be protected from destruction at the scientist's hand.

The stem cell debate often sounds as though everything boils down to status of the embryo. Is the preimplantation embryo outside a woman's body (*ex vivo*) a living person with inviolable dignity or not? If not, then the scientists may morally dismantle blastocysts and use the inner cell mass as a source of pluripotent stem cells for research and therapy. If the early embryo from fertilization and activation *ex vivo* is a full human person, however, then conservative Protestants and Vatican Catholics are right to demand that the research be shut down in order to protect the dignity of the unborn.

Is this the essence of the stem cell debate? Is this the only thing at stake? No, at least not in our judgment. While the question of the embryos status may be significant, only an inattentive and superficial rendering of the present situation would draw the picture in such bifurcated and simplistic terms. Yet, a more exhaustive and subtle understanding of what is at stake is difficult to find. Within each of the three different frameworks we discuss in this book, one can find two or more complementary if not competing and even irreconcilable positions. The ethical options are many.

This chapter looks at basic options from within the first bioethical framework: the *embryo protection framework*. Ethicists taking differing positions within the framework focus on the issue: Does the embryo at the blastocyst stage outside a woman's body require protection from destruction at the hands of laboratory researchers?

The battlefield in question is the derivation, not the benefits, of stem cells. The battle is over the moral status of the preimplantation embryo—the *ex vivo* blastocyst in the petri dish—from which human embryonic stem (hES) cells are derived. The central question of this first framework is this: Does the blastocyst have morally protectable dignity, so that we are forbidden to dismantle it when pursuing medical research? One may answer "yes" or "no" to this question. The very addressing of this question places the discussion within the embryo protection framework.

The position within this framework opposing stem cell research is most frequently associated with the Vatican, but is shared by many American evangelicals. Opposition to stem cells can be expected from the magazine *Christianity Today*, from denominations such as the Southern Baptists, as well as from conservative political advocacy groups such as the "Center for Bioethics and Culture" network and "First Do No Harm."[3] Because of the strength of the official Roman Catholic voice, this framework is often taken to be *the* religious framework. When this happens, opponents of this framework deride religion as antiscience. This is a big mistake.

The Structure of the Embryo Protection Framework

The National Council of the Churches of Christ in the USA (NCC) frames the ethical questions surrounding stem cell research this way: "As with the abortion debate, much of the stem cell debate turns on the differing views we hold on the moral status of human embryos."[4] Ethics and embryos belong together for the NCC; what we say about abortion applies to what we say about stem cells, or at least the central question is the same.

The moral proscription against hES cell research from within the embryo protection framework is characterized by several key assumptions and commitments. The first assumption is that the embryo, from the moment of conception, is a self-organizing entity with moral status. Sperm and egg are the property of the father and mother; but once fertilized the egg becomes an independent entity. It is a living being in its own right, a self-developing and self-maintaining unity under the governance of its own genomic plan. William B. Hurlbut says the "act of fertilization is a leap from zero to everything."[5]

The second assumption made by embryo protectionists is that a human life gets destroyed. To obtain embryonic stem cells, a blastocyst or early embryo at four to six days after activation needs to be disaggregated. Procurement of hES cells requires the destruction of an embryo, and this implies the destruction of a potential human person.

The third assumption is that a moral decision must be made regarding the moral status of the early embryo in the laboratory setting, *ex vivo*. Opponents of hES cell research argue that the embryo—whether *ex vivo* or *in vivo*—is of equal moral status to any other human being, and this forbids destruction. Because the intrinsic value of the blastocyst equals that of a living human person, stem cell opponents make their primary commitment to the protection of the embryo. Nigel Cameron of the Center for Bioethics and Culture rests his case on "human life with a dignity which is intrinsic and, therefore, with an inalienable moral standing." This permits extension of the abortion debate to early embryo research and other "issues of life and death *ex utero*."[6] Note the logic: the blastocyst has dignity; because it has dignity it is morally protectable; because it is morally protectable, its destruction in scientific experimentation constitutes murder. Stem cell harvesting from a blastocyst is akin to abortion.

To add rhetorical force, embryo protectionists may describe the embryo as the most vulnerable of human beings. Lutheran bioethicist Gilbert Meilaender is sympathetic to this position. "The embryo is, I believe, the weakest and least advantaged of our fellow human beings," says Meilaender and, citing Karl Barth adds, "and no community is 'really strong if it will not carry its . . . weakest members.'"[7] Concern for the weak and vulnerable among us—the poor, the sick, and the unborn—places embryo protection within a larger framework of love for those in need.

It is important to note that the embryo protection framework is characterized by the bioethical principle of non-maleficence—it ethically frames the stem cell debate as a matter of avoiding doing harm. Regardless of what good might result for future suffering persons who might benefit from today's research, the issue of doing harm or avoiding harm to the embryo takes precedence in the debate.

On what grounds might we think the early embryo possesses a dignity that forbids scientists from harming it? The most sophisticated account is provided by Vatican Catholics. We call it the *genomic novelty position*. This position, articulated already in the 1987 encyclical *Donum Vitae*, provides the foundational moral logic for what would later become the official Roman Catholic position on the stem cell debate.[8] *Donum Vitae* argues that three elements are crucial to the creation of a morally defensible human individual: the father's sperm, the mother's egg, and a divinely implanted soul. *Donum Vitae* notes that at fertilization a novel genetic code—neither that of the mother nor that of the father—is created. *Donum Vitae* takes this genomic novelty to be evidence of the presence of a unique individual, and thus reasonably the moment of ensoulment. Ensoulment is the event that establishes a divine moral claim, so that the destruction of the blastocyst constitutes not only murder but an offense against God's creation. Alleged empirical evidence that the early embryo has this divinely ascribed status is the uniqueness of the person-to-be's unique genetic code. Once a unique genome has been established, then it is morally incumbent on us to protect it from harm.

Donum Vitae was composed before Dolly the sheep was born. With the advent of cloning, this argument for genomic novelty becomes problematic. We now can create multiple embryos with the same genetic code. This is done routinely with farm animals. Application of cloning technology to humans is being considered. Would the existence of multiple persons with identical genomes challenge Vatican assumptions about the relationship of the soul to genetic uniqueness? Do twins have to split a single soul, with each getting only half a soul? Apart from being persons with identical genomes, what place do monozygotic twins—naturally born twins with identical genomes—have in this ensoulment scheme? If twins can each have their own soul, then why not clones? And if they do, then the argument from the position of genomic novelty slips through the fingers like sand.

The metaphysical foundation upon which the Vatican position is constructed is the belief in God's creation of a new spiritual soul that is imparted to the zygote at conception. The Christian tradition admits of different stages or levels of ensoulment: soul as initial animation, the capacity to reason, and then, finally, what our recent popes have called the "spiritual" soul, replete with reason and immortality. In effect, today's Vatican places the final stage, full spiritual ensoulment, at the very beginning, rejecting any version of delayed hominisation. Even if we may not know scientifically the exact moment infusion of a soul takes place, we can say philosophically that this event is precipitated by the establishment of the new genome. As a default position, from this perspective, for ethical reasons we must date ensoulment at conception, not later.

Because of the presence of the soul or even the readiness of the soul, the zygote has dignity. With dignity the early embryo may not be manipulated, compromised, or destroyed. But, this is not all. This soul is tied to only an individual human person, not to a group. Each person is unique in the eyes of God. By adding this second premise—that souls belong only to unique individuals—Roman Catholic moral theologians have seen the establishment of a new genome at conception as a way to identify a unique individual. The presence of an immortal soul plus a genetically unique individual have become two planks in the Vatican platform for building an argument against stem cell research. The early embryo at the blastocyst stage constitutes a human person in the fullest sense, and we may not make such a person a means for some further end such as medical research.

In 1995 the late Pope John Paul II published his widely read encyclical letter, *Evangelium Vitae* or "The Gospel of Life." In it he reiterated the commitment of *Donum Vitae* and other precedents, namely, that human life begins at conception and this warrants moral protection. Worried about abortion killing us on the front end and euthanasia—pulling the plug on the terminally ill—on the back end, the Holy Father warned the civilized world that it is drifting perilously close to becoming a "culture of death." Contraception, abortion, physician-directed suicide, are new practices that have unleashed a *"conspiracy against life."*[9] When President George W. Bush visited the Vatican in the spring of 2001, the pope extended application of the "culture of death" to include stem cell–researching scientists.

Vatican Variants

Three further variant positions within the embryo protection framework bear mentioning. One is a pair of *arguments from potentiality*. Distinguishing between a potential person and an actual person is the fare of the developmentalists. Even though the embryo will eventually become a person with protectable dignity, the establishment of the genome is not yet at that stage. At this point a moral argument could go one of two directions, one to permit stem cell research and the other to forbid it. Because the embryo is only a potential person and not an actual person, the destruction of the blastocyst to harvest pluripotent stem cells might be considered morally permissible. This variant can include respect or valuing of the early embryo, but not considering it beyond destruction.

Karen Lebacqz offers an argument that bears some similarities to such a view, though she rejects the language of "potential" and "actual" person. Lebacqz puts it this way: "First, the embryo or tissue must be valued. . . . To respect the embryo is to affirm that the value of the embryo or tissue is *not* dependent on its value for us or its usefulness to us. Respect sees a value in itself beyond usefulness. . . . Second, such an entity can be used in research and can even be killed. To do so is not in itself disrespectful."[10] Lebacqz respects the *ex vivo* blastocyst for its humanity; yet she finds it ethical to disaggregate it for stem cell research, because respect for humanity requires different actions at different stages of life. For example, respect for adults requires that we obtain their informed consent before treating them medically; respect for children does not require their informed consent, because they do not yet have the capacity to consent.

Thomas Shannon offers another variant, utilizing scholastic philosophy to support a developmentalist position. Right after fertilization, the zygote is a living entity, to be sure, and it possesses human nature. But this is a *common* human nature with an array of potentials. Even if personhood is one of the potentials, the activated zygote is not yet an individuated person; thus it does not yet have full human dignity. Embryogenesis is a process, and dignity cannot be applied until we have an individual person, which comes only well beyond the blastocyst stage of development. "Persons," Shannon argues, "have a dignity; natures have a value. The dignity of the person grounds a more absolute standing. . . . The value of human nature does not generate the same level of protection. . . . Nonetheless, it is human nature and it is to be valued."[11] In sum, one can hold a position that affirms an early embryo is a person in potential and, with an appropriate level of respect, still support embryonic stem cell research.

Meilaender, whom we associated with the stricter embryo protection position above, finds such arguments for respect or value ascribed to the predifferentiated embryo less than convincing. He complains that, "If we forge ahead with embryonic stem cell research, we simply scrap the language of respect or profound respect for those embryos that we create and discard according to our purposes. Such language does not train us to think seriously about the choices we are making, and it is, in any case, not likely to be believed."[12]

Meilaender is among those who hold the parallel yet contrary position from that of Lebacqz and Shannon. Meilaender argues that the embryo—though obviously not a human being in the full sense—is still at minimum a *potential* human being; and this potentiality warrants protection. The blastocyst should not be treated as a means to some further end; to do so would be to ignore the continuous development of the individual from the embryonic to fetal and infant stages. It follows that stem cell research should be halted.

In sum, to view the *ex vivo* blastocyst as a potential human person can lead down two separate moral paths. The sign on one path reads: Because this is a potential person and not yet an actual person, disaggregation for stem cell research is permitted. The sign on the other path reads: Even though this is not an actual person, it is still a potential person; therefore, it would be immoral to destroy it to further the interests of scientific research.

A variant of this variant we call the *better-safe-than-sorry position*. Differing from the *Donum Vitae* Catholic position, better-safe-than-sorry advocates do not make the assumption that we know the moment when personhood or morally defensible dignity begins. Rather, they recognize that an early embryo could be considered a potential person; but the threshold at which it crosses into actual personhood remains unspecifiable. With this in mind, the better-safe-than-sorry position holds that, because we cannot definitively identify the moment when developing life becomes morally defensible, we must take the most morally conservative position possible vis-à-vis the embryo. The earliest possible moment becomes the safest. By default, protection of the individual begins with protection of the fertilized egg.[13]

Another variant position might be called the *discarded embryo position* or the *nothing-is-lost position*. This view depends on the fact that, to date, the preponderance of embryonic stem cell research has been conducted on "excess" embryos originally created for purposes of IVF. If not placed into a woman's uterus, they are frozen and eventually discarded. In the United States and United Kingdom thousands of such embryos exist in storage freezers. Those who hold the discarded embryo position believe it is morally licit to use for research embryos that will otherwise be destroyed. What is illicit is the deliberate creation of embryos that will be destroyed for research purposes.

The discarded embryo position finds voice in the conversation as the nothing-is-lost position. This position is invoked when we distinguish between using existing fertilized ova that have been discarded by IVF clinics and the deliberate creation of new embryos to establish new cell lines. The nothing-is-lost position appeals to *exempting conditions* to destroying human life such as (a) observing that existing embryos will be discarded anyway and (b) observing that as research material they could be indirectly life-saving. Gene Outka draws this conclusion: "The creation of embryos for research purposes only should be resisted, yet research on 'excess' embryos is permissible."[14]

A bioethicist who relies upon precedents drawn from the Orthodox Christian tradition, H. Tristram Engelhardt, Jr., agrees that the use of discarded embryos stored in IVF clinics is morally licit, because such use draws something good out of an otherwise immoral situation. "There is no bar in principle against using for a good end something that has been acquired by heinous means, as long as one has not been involved in (1) employing these evil means, (2) encouraging their use, (3) avoiding their condemnation, or (4) giving scandal through their use. One can drink water from a well that was dug by unjustly forced labor."[15]

These theological deliberations have secular counterparts. What is becoming the accepted ethic for stem cell research is approval of use of discarded IVF embryos accompanied by disapproval of producing fresh embryos and then destroying them at the blastocyst stage. Whether in the UK, Japan, Singapore, or the United States, preference for embryo derivation is given to those fertilized ova initially created for the purposes of fertility treatment—for a purpose other than scientific research—but, then, no longer planned for fertility use and slated to be discarded. Research embryos must be created for some other purpose and then diverted into research. If one asks why, no scientific reasoning can explain such a policy. There is no scientific reason why discarded IVF embryos should be permitted while newly created ones are discouraged or even forbidden. What we see here is religious reasoning infiltrating scientific research standards and secular public policy.

Some Roman Catholics counter the discarded embryo view in an extreme way. The extremists contend that laboratory use of excess or discarded IVF embryos makes the scientist complicit in the crime of embryo destruction. Richard Doerflinger, spokesperson for the U.S. Conference of Catholic Bishops, holds this extreme view. He says, "Intentional destruction of innocent human life at any stage is inherently evil, and no good consequence can mitigate that evil."[16]

In summary, here are the positions within the embryo protection framework just discussed.

- Permit research because the blastocyst has not yet been conferred dignity
- Stop research because full moral dignity was given by God at conception
- The *ex vivo* blastocyst is a potential person. It follows that . . .

  (1) the early embryo has "respect" but not "dignity," therefore, permit research
  (2) the early embryo has morally protectable "dignity," therefore, stop research

- Better-safe-than-sorry, therefore, stop research
- Discarded embryo/nothing is lost, therefore, permit research

What Does Nature Say about the Early Embryo?

Michael D. West reports that he is the target of a religiously motivated firing squad. The rhetoric has been explosive. Richard Land of the Southern Baptist Convention called therapeutic cloning, "high-tech cannibalism." One headline in another magazine read: "Dr. West and Bin Laden: Cloning and Terrorism Are Both Clear and Present Dangers."[17] West opines: "Inflammatory rhetoric distorts the debate."[18] Yet, West too can rise to the rhetorical occasion: "Some members of the religious establishment have . . . attacked scientists such as me as the authors of the 'culture of death.' I see it from the opposite perspective—those who would deny a dying patient lifesaving therapy on the basis of obscure theological speculation are the culture who promote death, however unintentionally."[19]

Beneath the rhetoric we need to ask: Are stem cell scientists really baby killers? Arguments developed solely within the embryo protection framework make the question of the embryo's status unavoidable.

What does West himself think about the matter? He tries to analyze the sixth commandment, "Thou shalt not kill." He asks whether it refers to the destruction of the blastocyst in the petri dish. Is it murder? "I believe the answer is no," he writes, "for it is not the taking of an actual human life." He goes on to report that the Christian Bible, right along with the Torah and Koran, does not provide clear instructions regarding the preimplantation embryo. They could not of course, because what we are talking about is the fruit of modern science. "Ancient dogmas provide little of concrete use in formulating modern standards of ethics and law relating to the pre-implantation embryo."[20] It appears that the ethicist will have to engage in the practice of discernment, that is, trying to apply ancient commitments to modern problems.

West then raises the question we all want to ask: Does life begin at conception? No, he answers again, certainly not individual human life. "The sperm and egg cells are very much alive," to be sure. But does their coming together in conception create what we know as an individual human life? No, because no individual is established prior to implantation in a woman's body somewhere around the 14th day. At this early stage, "quite often a single fertilized egg splits to make twins. Since they come from a single fertilized egg, they share all the same genes. . . . Certainly in the case of these twins, no one can assert that they were individual human beings at conception."[21]

Individualization "occurs at about fourteen days after fertilization. Up to this point there has been no movement toward building a human body. The cells in preimplantation embryos contain no body cells of any kind, not even any cells on their way to becoming body cells. They are blank embryonic and immortal stem cells." What happens when the embryo becomes connected to the mother's uterine wall becomes decisive. At this point the primitive streak appears, the precursor to the backbone. Only at this point does nature decide whether we will have one person or two or more. "The primitive streak is therefore a very useful line drawn in the dust on the ground from which we are made, drawn not by arbitrary human convention but by nature itself."[22]

This observation regarding early embryo development is relevant to the Vatican argument of *genomic novelty* outlined above. The Vatican reveres genomic novelty much more highly than nature does, so it seems. Prior to the appearance of the primitive streak, nature can deliver twins, quadruplets, even octuplets. Only when the relationship of the embryo to the mother is established does nature decide how many children will share the genome established earlier at conception. What this implies for the Vatican argument is this: If the immortal soul is connected to an individual person, then ensoulment would have to be dated at day 14 and not at conception. Further, it is misleading to tie human individuality so closely to our genetic code.

Is It Abortion, Really?

This observation regarding the first appearance of individuation with the primitive streak at 14 days offers us an opportunity to ask the question: Does the harvesting of stem cells constitute abortion? The three coauthors of this book think the answer is no.

We would define elective abortion as the surgical removal of a fetus from a woman's body. Stem cell research takes place outside a woman's body, not within it. Furthermore, it takes place at a stage of embryo development prior to implantation and the establishment of a single individual human life. In most cases, a woman cannot actually detect the presence of the *conceptus* or even the early fetus until many weeks after implantation. Abortion decisions arise weeks or months beyond the period of time in which embryo research takes place.

Another distinction is important. The focus of opposition between pro-choice and pro-life arguments is the woman's right to choose. This argument over the right to choose arose long before the appearance of stem cell research and when quite different cultural issues were being debated. The abortion controversy raging in North America since the 1960s centers on the question: Does a woman have a right to determine what happens to her own body? Does this right include the choice to decide whether to keep and care for a fetus or ask that it be surgically removed in a medically safe environment? Should elective abortions be legal? Pro-choice proponents say yes, while those usually called "pro-life" say no. Neither set of proponents believes that destruction of a fetus is a good thing. Nobody advocates the willy-nilly practice of abortion. Even our two coauthors who defend a woman's right to choose see themselves as pro-life in a fundamental sense.

As mentioned earlier, the three authors of this book take somewhat different positions on elective abortion. One sympathizes with what is usually called the pro-life position, while the other two are pro-choice. All three of us are pro-stem cells. This can happen because the issues surrounding elective abortion are not the same as those involving stem cells. In our judgment, it is misleading to equate the stem cell controversy with the abortion controversy.

One more observation. When embryo protectionists ascribe the status of an unborn human person to the laboratory blastocyst, they grant it a worth from which they draw moral direction. This tacitly treats the blastocyst as sacred. It makes the early embryo an end in itself; and it subordinates other ethical concerns to it. One might understand and even embrace this logic within the abortion debate if one attributes personhood to the unborn. But, a pluripotent or even totipotent stem cell is not in itself an unborn baby. It is not sacred. It has its own worth, to be sure, but at the blastocyst stage it is disingenuous to suggest that it has the worth or dignity of an individuated human being. It is a mistake to treat a cell as sacred, to treat a blastocyst as if it were an individual person with dignity. We will say more on this later.

Objections?

Among the members of various theological communities who establish positions regarding bioethical issues in general, let alone stem cell research in particular, Roman Catholic theorists have exerted the greatest amount of energy and made the most thorough use of intellectual resources. For this reason, we believe Roman Catholic arguments deserve detailed attention. Yet, the Vatican's strong and well-argued stand makes Catholics more vulnerable than others to criticism. Jean-Pierre Changeux, professor of neurobiology at the College de France, referencing reproductive ethics, gives voice to a prevalent attitude among scientists. "Certain positions taken by the Catholic church are a source of concern to a citizen-scientist who sees himself, as I do, as both tolerant and responsible . . . science exists to help men and women survive illness and to live better lives. But the positions adopted by the Vatican in the name of moral truth are sometimes contrary to what one would suppose morality to be."[23] Note that this scientist objects to the moral positions taken by the Vatican. The Vatican's moral positions are not in themselves antiscience. Rather, the Vatican affirms science while seeking to guide science through ethics.

What we have tried to show here is that the moral judgments rendered by Vatican moral theologians are constructed from within the embryo protection framework. One might disagree with the Vatican from within the same ethical framework; or one might make an alternative judgment based upon reasoning in one of the two other ethical frameworks. In the chapters that follow we will show what many of the alternatives look like.

Conclusion

Even though the Vatican holds a prominent place among those who oppose stem cell research, it would be a mistake to believe that Roman Catholics are the only opponents or that the official Roman Catholic position is the only position held by those who oppose stem cell research. It would also be a mistake to think that all Roman Catholic moral theologians oppose stem cell research, despite the strong stand of the Vatican. Even some Roman Catholic ethicists who give full support to morally protectable dignity for the fetus and the early embryo find they can support hES research in good conscience.

We have tried to show in this chapter that one or more moral positions can be argued from within the embryo protection framework.

Still more can be said regarding diversity of moral commitments. Importantly, many who hold to the dignity of early embryos also hold to other significant values, such as social justice. Margaret A. Farley speaks for many of her fellow travelers when she reminds us that "Roman Catholics . . . tend to worry, like people in other traditions, about issues of justice, ecology, and the well-being of the whole Earth. Apart from the moral status of the embryo, Catholic concerns are focused on questions of equity in the shared lives of people across the world."[24] Farley, along with others in non-Catholic traditions, tries to bind together her full respect for the dignity of the early embryo with her commitment to the well-being of all persons and even the well-being of our planet.

Because these broader concerns play an important role yet do not fit within the embryo protection framework, other more appropriate frameworks are needed. Concerns for justice and planetary well-being, for example, are most frequently articulated within the future wholeness framework, which we will describe later. In the meantime, we turn to the framework in which protection of human nature becomes the focus of ethical concern.

Notes

1. Tony Reichhardt, "Studies of Faith," *Nature* 432 (2004): 666.

2. Ann Coulter, *The Church of Liberalism: Godless* (New York: Crown Forum, 2006), 194.

3. Center for Bioethics and Culture Network (CBC): www.thecbc.org; First Do No Harm: www.donoharm.org.uk/.

4. The National Council of the Churches of Christ in the USA, "Fearfully and Wonderfully Made: A Policy on Human Biotechnologies," www.ncccusa.org/pdfs/BioTechPolicy.pdf, lines 307–8.

5. William B. Hurlbut, "Altered Nuclear Transfer as a Way Forward for Stem Cell Research: Moral Reasoning and Scientific Evidence," unpublished paper delivered at the International Seminar on "Dilemma of the Stem Cell: Research, Future, and Ethical Challenges," Islamic Organization for Medical Sciences at the World Health Organization, Cairo, Egypt, November 3–5, 2007. This assumption is based upon biological observation, not upon a theological assertion regarding ensoulment. The ensoulment argument adds another layer to the embryo protection argument, as we shall see.

6. Nigel M. de S. Cameron, *The New Medicine: Life and Death after Hippocrates* (Chicago, IL: Bioethics Press, 1991, 2001), 100–101.

7. Gilbert Meilaender, "Some Protestant Reflections," *The Human Embryonic Stem Cell Debate*, ed. Suzanne Holland, Karen Lebacqz, and Laurie Zoloth (Cambridge, MA: MIT Press, 2001), 142.

8. Congregation for the Doctrine of the Faith, *Instruction on Respect for Human Life in Its Origins and on the Dignity of Procreation (Donum Vitae)* (February 22, 1987), Acta Apsotolicae Sedis 1988, 80, 70–102. See also John Paul II, *Evangelium Vitae* (March 25, 1995), Acta Apostolicae Sedis 1995, 87, 401–522.

9. Pope John Paul II, *The Gospel of Life* (New York: Random House, Times Books, 1995), 22.

10. Karen Lebacqz, "On the Elusive Nature of Respect," *The Human Embryonic Stem Cell Debate*, ed. Suzanne Holland, Karen Lebacqz, and Laurie Zoloth (Cambridge, MA: MIT Press, 2001), 159.

11. Thomas A. Shannon, "Grounding Human Dignity," *Dialog* 43.2 (Summer 2004), 117.

12. Meilaender, "Some Protestant Reflections," 145.

13. This position was classically articulated by the Vatican in a 1974 document entitled "Declaration on procured abortion, 18 November 1974," located online at: www.vatican.va.

14. Gene Outka, "The Ethics of Human Stem Cell Research," *God and the Embryo: Religious Voices on Stem Cells and Cloning*, ed. Brent Waters and Ronald Cole-Turner (Washington, DC: Georgetown University Press, 2000).

15. H. Tristram Engelhardt, Jr., *The Foundations of Christian Bioethics* (Lisse: Swets and Zeitlinger, 2000), 261.

16. Richard Doerflinger, "The Policy and Politics of Embryonic Stem Cell Research," *National Catholic Bioethics Quarterly* 1.2 (Summer 2001): 143.

17. Michael D. West, *The Immortal Cell: One Scientist's Quest to Solve the Mystery of Human Aging* (New York: Random House/Doubleday, 2003), 207.

18. West, *Immortal Cell*, 219.

19. West, *Immortal Cell*, 231.

20. West, *Immortal Cell*, 212–13.

21. West, *Immortal Cell*, 213–14.

22. West, *Immortal Cell*, 214–15.

23. Jean-Pierre Changeux and Paul Ricoeur, *What Makes Us Think?* (Princeton, NJ: Princeton University Press, 2000), 260.

24. Margaret A. Farley, "Stem Cell Research: Religious Considerations," in *Handbook of Embryonic Stem Cells*, ed. R. Lanza, J. Gearhart, et al., vol. 1 (Boston, MA: Elsevier/Academic Press, 2004), 765–73.

CHAPTER SIX

∼

The Human Protection Framework

The second prominent ethical framework we call the *human protection framework*. The framework could easily be called *nature protection* or *anti-brave new world* or *anti-playing God*. The ethical goal is to defend what is "truly human" from violation by what is artificial, namely, genetic technology. The vocabulary includes frequent references to Prometheus, Frankenstein, pride, hubris, and the posthuman future. Critics of genetic science working within this framework frequently build their arguments on a floor of fear: the fear that the promised benefits of biotechnology offered by scientists and their paid ethicists covers over the risk of sacrificing what is precious about humanity, driving us toward a posthuman future. Human nature is threatened and we must protect it.

The impetus here is to protect our humanity against the dehumanizing forces of technology. Because almost nobody openly advocates creating a "brave new world" as depicted in Aldous Huxley's novel, those operating from within this framework usually oppose stem cell research. Yet, those who feel they are opposing something need to identify that which they oppose. The anti-brave new world rebels feel that they stand in opposition to an implicit worldview—a dominant or prevailing worldview—against which they need to take a stand. Lutheran theologian Philip Hefner describes this worldview. "We approach nature—including our own human nature—in terms of what we can make of it; nature is not something we accept, rather it is the object of our fantastic ability to reshape our world. Further, this characteristic is not incidental, but it reveals itself as a fundamental American habit of the heart."[1] The American habit that opponents to a brave new world shun is the habit of reshaping what nature has given us. To technologize and transform the human race into a posthuman future strikes fear into the hearts of those who appreciate nature as it is.

Fear of an approaching brave new world leads moralists to take as their point of departure defense against potential undesirable future consequences of biotechnological research generally, embryonic stem cell research specifically. This framework has been most coherently articulated by the majority members of President George W. Bush's Council on Bioethics.[2] Resonating with the moral warnings of Huxley's famous *Brave New World*, this framework comprehends the stem cell debate by first imagining undesirable and unforeseen consequences of stem cell research, and working backward from these undesirable futures to regulate present-day policy. Opponents of stem cell research employing this framework usually assume a reverence for nature, for our human nature as we have inherited it from our evolutionary past. They enjoin us to accept our nature as it is, complete with our vulnerability to disease and our "natural" limits to reproductive options. They fear getting too far away from what is natural.

The mood of the human protection framework is expressed forcefully by Leon Kass, University of Chicago professor and former chair of President George W. Bush's Council on Bioethics. When denouncing reproductive cloning, Kass said: "I exaggerate somewhat, but in the direction of truth: we are compelled to decide nothing less than whether human procreation is going to remain human, whether children are going to be made to order rather than begotten, and whether we wish to say yes in principle to the road that leads to the designer hell of *Brave New World*."[3] Kass's denunciation of reproductive cloning here leads indirectly to a correlate denunciation of therapeutic cloning and thereby a ban against embryonic stem cell research. The technique for therapeutic cloning could be co-opted by those who would use it for reproductive cloning. "What we need is an all-out ban on human cloning, including the creation of embryonic clones."[4]

Opposition to Playing God

Three weight-bearing pillars hold up the structure of the human protection position: "anti-playing God," the "wisdom of repugnance," and "neonaturalism." Let us look at them in turn.

The first pillar supporting the human protection framework is a new commandment: Thou shalt not play God. Opponents to playing God argue that while any individual case of technoscientific advance might be ethically permissible, the cumulative effects of widespread biotechnological pursuit cannot be adequately anticipated. Human nature protectionists worry that human desire for technological advancement will go unbridled, such that genetic medicine will become genetic enhancement, ultimately producing an unjust society of genetic haves and have-nots. As such, they argue that it is pride or *hubris* to believe that through science and technology we can control human biological destiny.

Emphasizing the virtues of life as it is given by nature, anti-playing God moralists warn that pursuit of biotechnological advance degrades the value of what is natural generally and of human nature particularly. They warn us against "playing God" through scientific intervention into natural processes. They warn us against Frankenstein, against Promethean *hubris* that might lead our scientists to violate nature, because nature would then retaliate with destructive force. Leon Kass shrinks away in horror from "the Frankensteinian hubris to create human life and increasingly to control its destiny; man playing God."[5] Within this framework some ethicists conclude that destruction of embryos, even in pursuit of medical benefits, risks coarsening society to the value of nascent human life; its future risks outweigh its present benefits.

In the ancient world of the Greeks, people feared the gods of Olympus. They sought to propitiate the gods in order to ensure health and prosperity. We moderns no longer believe in such gods. In their place we have set nature. Instead of revering the gods, some among us believe we should revere nature. We have replaced an avoidance of acting *contra deum* (against God's will) with a new proscription against acting *contra naturam* (against nature). To avoid "playing God" means for us to avoid acting contrary to nature. What has happened is that nature has become tacitly sacred.

Implicit in the anti-playing God position, therefore, is a reverence for nature that makes nature functionally sacred. DNA has become an icon of nature's sacredness. Frequently treated as though it were the essence or secret to life, the genetic code is lifted up by this position as quasi inviolable. With their secular sacrality, the genes should be morally off-limits to science. Scientists who alter the human genome are accused of playing God, of crossing the line that led to the punishment of Prometheus by Zeus. This sense of the sacred in our genome leads directly into the third variant, neonaturalism.[6] But first, the wisdom of repugnance.

Ethics Built on Yuck

The second pillar holding up the human protection framework is the "wisdom of repugnance." The "yuck factor" is the term we prefer. When the cloning controversy broke out at the announcement of Dolly the sheep in late February 1997, the entire world went, "Yuck!" How dare scientists manipulate and violate our sacred DNA! The reaction was visceral. Reproductive cloning was immediately perceived to be a threat to human individuality, identity, and dignity. Nobody wanted it. So, scientists were blamed for presenting us with this threat, a threat nobody welcomed.

On the cover of *Der Spiegel*, a German magazine akin to the English language magazine *Time*, we saw a picture of Dolly the sheep, a line of marching Adolf Hitlers, a line of marching Albert Einsteins, and a line of marching Claudia Schiffers. In large capital letters stood out a theological phrase, *Der Sündenfall*—that is, "the fall into sin." Our mad scientists, suggested *Der Spiegel*, were causing the entire human race to fall into sin. This visceral repudiation of the new technology we label "yuck." Yuck indicates an amorphous and unarticulated feeling, yet a very strong feeling, to be sure.

Leon Kass refers to "yuck" as "repugnance."[7] Furthermore, Kass believes repugnance tells us which direction our ethics should follow. The Kass position grounds ethics in the recognition that advancing technologies can strike with visceral repugnance that belies any easy rational articulation. This repugnance functions as a moral alarm, alerting us to the potential harms of "unnatural" intervention.[8] "*Offensive, grotesque, revolting, repugnant*, and *repulsive*" are the words Kass lifts up as reactions to genetic technology, specifically, cloning. Such words count methodologically for Kass, because he relies upon the wisdom inherent in the emotional intuition of repugnance. "Repugnance is the emotional expression of deep wisdom, beyond reason's power fully to articulate it."[9] Genetic technoscience dehumanizes us, because it alienates us from our sense of belonging to nature.

Kass fears the slippery slope. He fears that we might go "from 'Yuck' to 'Oh?' to 'Gee Whiz' to 'Why not?'"[10] So, to ward off the advancing attack of the technoscientists, Kass wants to return to the initial yuck factor and construct his ethics on the wisdom of repugnance.

The Kass position seems irrational, and Kass admits it. The scientists who gave us cloning and now stem cells are very rational, and if this is what rationality consists of, then Kass wants nothing to do with it. So, he appeals to what is prerational, to the wisdom of repugnance. Recall his basic premise: Repugnance is the emotional expression of deep wisdom, beyond reason's power completely to articulate it.[11] Yet, Kass is aware of the dangers of a total abandonment of reason in ethics. Once he has taken his stand against reason in laying the foundation for ethics, he moves back somewhat toward a compromising middle. Once the prerational foundation is laid, Kass would adopt the reasoning process for drawing out ethical implications.

The New Naturalism

What is it that is revealed by our deep wisdom as it surfaces in the emotion of repugnance? Kass answers: our humanity, our fundamental human nature. This leads to the third pillar holding up the moral judgment against genetic research within the human protection framework, neonaturalism.

In Kass's work, neonaturalism takes the form of reverence for humanity as we find it, as nature has evolved it. We human beings are natural creatures, and the advance of technology threatens to denaturalize us. Rather than seek transformation through technology, especially biotechnology, we should enhance our appreciation for our finitude and our natural limits. In particular, Kass is concerned about that dimension of human nature involved in procreation: sex, marital fidelity, giving birth to children, and caring for children. It is this natural state that comes to the surface when we feel yuck, when repugnance expresses deep wisdom.

Few contemporary thinkers can match Kass's eloquence when reminding us of truths that lie deep and only semiarticulate within us. We must not avoid the "concrete existential questions surrounding birth and death, sickness and health, suffering and flourishing," he writes. We must reach down "to the central concerns of human life: identity and individuality, freedom and finitude, embodiment and selfhood, sexuality and procreation, and the deeply mysterious longings of the human soul."[12]

The appearance of the "soul" in Kass's list of concerns reminds us that neonaturalism comes in both religious and secular forms. Both are trying to protect our humanity from the perceived threat of biotechnology. Among the religious variants, some Roman Catholic natural law theorists worry that any experimentation with early embryos will separate sexuality from reproduction and violate the traditional integrity of family life. Engineering the genetic code of a child is tantamount to refusing to accept a child as a gift; and destroying a potential child in a petri dish to forestall the consequences of genetic disease testifies to our refusal to accept the hand that nature has dealt us. The root problem is our attempt to control nature. Some Catholics argue that to let nature take its course is to yield to the will of God. To engineer nature is to resist the will of God.

The human protection framework focuses on reproduction. Kass and others contend that babies should be born without the use of technologies. And parents should accept what nature gives them. This is why reproductive cloning becomes such a hot target. This focus on reproduction and technology extends to research that uses embryos, including embryonic stem cell research. Any research with embryos risks compromising the natural value of reproduction. Stem cell research can be dehumanizing when it coarsens us to the wonder and intrinsic worth of reproduction.

What is natural and what is most worth protecting in the face of techno-logical advance is the biology of family life. Opposition to stem cell research from within the human protection framework is indirectly an endorsement of traditional family life. The rejection of blastocyst destruction in stem cell research is connected to the rejection of cloning. It is based upon the desire to protect the natural processes of human embryogenesis and development. Research and technologies that disrupt these natural processes at any stage risk the Frankensteinian hubris the human protectionists so abhor.

We find this fallacious. Many philosophers call this form of reasoning the *naturalistic fallacy*, because it argues without warrant from *what is* to *what ought to be*. It assumes that what we have inherited from our natural past provides the values that should ethically orient our future. We, the authors of this book, disagree. We think that orienting commitments should be drawn from God's promise for a fulfilled future, not an idealized past. For us, ethics looks forward to transformation, not backward to what is given in nature.

The Kass argument is essentially a secular naturalist argument. Variants can also be found among Lutherans and Roman Catholics. One can dis-cern a hint of such theological naturalism in Gilbert Meilaender. Lutheran Meilaender wants us to receive each child as a gift and honor the biological structures of traditional family life. Meilaender, who served with Kass on the president's Council on Bioethics, further wants us to respect our bodies as na-ture has bequeathed them to us. This is because God, by becoming incarnate, has affirmed the goodness of our bodies. Too much technology applied to improving our bodies risks separating us from our natural origins. "Eschewing these technologies," says Meilaender, is a way of "honoring the limits of that bodily life which God has taken into his own life."[13]

A theologically subtle and scientifically sophisticated version of the Ro-man Catholic form of neonaturalism is found in the work of British biologist and theologian Celia E. Deane-Drummond. Like Kass, she appeals to wis-dom. For Deane-Drummond, wisdom comes from both the Bible and human intuition. Armed with wisdom, she marches forward. "We need to challenge two opposing mythologies. The first mythology is that living beings are just their genes, and cloning in this scenario represents a threat. The second mythology is that it does not really matter what manipulations are made, since living beings are more than their genes. An ethic of wisdom seeks to adjudicate when and if such changes can be made, given the context of de-veloping an attitude of love and respect towards all creatures."[14] Her appeal to wisdom leads her cautiously toward the discarded embryo position within the embryo protection framework. "This position tentatively would support a very restricted use of already existing spare IVF embryos in order to create a limited number of embryonic stem cell lines."[15]

Returning to secular neonaturalism, we find in Jeremy Rifkin a full-blown naturalism that seeks a "resacralization" of nature. Rifkin objects to "the upgrading of existing organisms and the design of wholly new ones with the intent of perfecting their performance." To protect our genetic nature from human technoscientific intervention and modification, he trumpets: "The resacralization of nature stands before us as the great mission of the coming age."[16] Rifkin claims that nature is good as we find it. Biotechnical interventions, such as embryonic stem cell research and regenerative medicine built on genetic modification, are violations of the moral order of nature. The ethical import of resacralizing nature is to provide moral warrant for preventing scientists from reengineering DNA.

From Immortal Cells to Immortal People

Perhaps Rifkin's nightmare would be biologist Stanley Shostak who writes that "a high priority should be placed on manipulating genes, fulfilling biotech's potential for creating a healthier and happier humanity."[17] Shostak, like West, is pursuing not simply immortal cell lines but also immortal people. Shostak believes a combination of cloning and stem cell research could beat the problem of death. "Eliminating old age and death is now within reach."[18] Shostak is a clear example of Promethean hubris, arguing on behalf of genetic advance from within the human protection framework. Yet, even with such a Promethean ambition, it is not necessarily the case that all ethical cautions must be thrown to the wind.

Perhaps it should not surprise us that stem cell researcher Michael D. West shares much with these neonaturalists. On the one hand, he strongly supports scientific advance in the direction of improving human health and well-being. "I actually value the title of Aldous Huxley's novel *Brave New World*, for we need to make a new world—not the world of Aldous Huxley's novel, but a world free of the scourges of diabetes and heart disease."[19] Yet, on the other hand, West opposes genetic enhancement and, thereby, opposes a posthuman future. "My nightmare is that the hubris of some scientists would have us engineering *people* from genetically modified cells. Their goal is 'enhancement,' to make 'superpeople,' individuals better than any living on the planet today."[20] Making superpeople is West's nightmare, not dream. Like others, he recognizes how what we used to see as *contra deum*—as against God's will—is today formulated *contra naturam*—against nature. With this in mind, he shares the "fear that, in the end, biotechnology will cause us in some way to lose our humanity."[21] Although there may be a threat to our humanity lurking in genetic science, West is convinced that the threat is not coming from human embryonic stem cell research. West shows that one can affirm stem cell and related genetic research while considering moral arguments from within the human protection framework.

Alternative Views of the "Truly Human"

The human protection framework is constructed in response to the question: what is "truly human" about human nature, and how might biotechnology compromise it? To date the answers given to this question have largely led to the denunciation of embryonic stem cell research. Is such a negative response inevitable, or might we imagine an understanding of the "truly human" that recommends *support* for stem cell research?

Many theologians have argued that humans are by nature technical beings. To shape and reshape our world is basic to who we are. Lutheran theologian Philip Hefner, for example, describes humans as "created co-creators" with God.[22] This view is connected to the theological notion of *creatio continua* (continuing creation) according to which God not only brought the world into being, but is continuing to shape the world in movement toward its completion and perfection. To be truly human in this case is to join with God in an ongoing creation of the world. Such a view resonates with the Jewish principal of *Tikkun Olam*, the command to join God in healing and completing the world. Both human and nonhuman nature can be seen as dynamic and evolving and not static and primordial.

If human nature was not "fixed" once and for all, if what it means to be truly human continues to develop over time, how does this change our framing of stem cell research? Might we support stem cell research because new technologies make us more distinctively human? Perhaps such an unqualified conclusion would not be warranted. To quote *Donum Vitae*, "what is technically possible is not for that very reason morally admissible."[23] On the other hand, when assessing what does count as morally admissible, we should keep in view the creative aspects of human nature. Stem cell research can be viewed as an expression of our capacity to shape and reshape the world in response to the call to reduce suffering and bring about justice, joy, and fullness of life.

Human Protection Meets Embryo Protection

It is worth noting that, like the embryo protection framework, the human protection framework portrays the stem cell debate in terms of non-maleficence. Accordingly, our primary moral responsibility is to guard against the potential negative consequences of biotechnological research. In this framework ethicists wish to avoid doing harm to our culture. The embryo protectionists, on the other hand, wish to avoid doing harm to the blastocyst.

In summary, the human protection framework is an application of a broad culturally articulated fear that advancing technology risks divorcing our consciousness as human beings from our biological embeddedness in the natural realm. More frequently than not, positions taken within this framework partner themselves with the embryo protectionists.

Arguments arising from within the human protection framework are not typically aimed directly at the stem cell controversy. Although vitriolic in opposition to reproductive cloning, judgments regarding stem cell research get swept into implied arguments against therapeutic cloning.

In this chapter we have identified three pillars holding up the framework for a defense against the encroachment of a "brave new world": the "wisdom—of repugnance" pillar, the "anti-playing God" pillar, and "neonaturalism" in both its theological and secular varieties. The human protection framework is a broad one, leading to a conservationist stand against the threat of technology in general. It seeks to prevent movement toward what some call the "posthuman future." Leon Kass, Celia Deane-Drummond, and Gilbert Meilaender want a human future, not a posthuman future. For this reason, they take a strong stand against reproductive cloning. Once this is clear, they take a weaker stand against human stem cell research. Deane-Drummond right along with Michael D. West can approve of stem cell research under controlled conditions.

The key to unlock the logic of this position is this: Does our human nature exclude or include our technological capacity? On the one hand, if we assume that by "human nature" we refer exclusively to what the human race has inherited from its evolutionary past, and if we want to protect this biological inheritance, then we would put up a stop sign to prevent technological innovation. On the other hand, if we assume that by "human nature" we include the fact that our species engages relentlessly in creative innovation and technological invention, then it would not surprise us that we give creative attention to our own DNA. The task of the ethicist in this latter case would be to guide creative innovation, not outlaw it.

Notes

1. Philip Hefner, "The Genetic 'Fix': Challenge to Christian Faith and Community," *Genetic Testing and Screening*, ed. Roger A. Willer (Minneapolis, MN: Kirk House, 1998), 76.

2. See President's Council on Bioethics, *Monitoring Stem Cell Research* (2004) and *Beyond Therapy: Biotechnology and the Pursuit of Happiness* (2004) located online at www.bioethics.gov.

3. Leon R. Kass, *Life, Liberty and the Defense of Dignity: The Challenge for Bioethics* (San Francisco, CA: Encounter Books, 2002), 146.

4. Kass, *Life, Liberty and the Defense of Dignity*, 167.

5. Leon R. Kass and James Q. Wilson, *The Ethics of Human Cloning* (Washington, DC: AEI Press, 1998), 18.

6. For a broad analysis of the fear of playing God in genetic research, see Ted Peters, *Playing God? Genetic Determinism and Human Freedom* (New York: Routledge, 2002). Other treatments include Andrew Dutney, *Playing God: Ethics and Faith* (Melbourne, Australia: HarperCollins, 2001) and John H. Evans, *Playing God? Human Genetic Engineering and the Rationalization of Public Bioethical Debate* (Chicago: University of Chicago Press, 2002).

7. Leon R. Kass, "Preventing a Brave New World: Why We Should Ban Human Cloning Now," *New Republic* 224.21 (2000): 30–39.

8. Kass and Wilson, *The Ethics of Human Cloning*, 17.

9. Kass and Wilson, *The Ethics of Human Cloning*, 18.

10. Kass, *Life, Liberty and the Defense of Dignity*, 146.

11. Kass, *Life, Liberty and the Defense of Dignity*, 150.

12. Kass, *Life, Liberty and the Defense of Dignity*, 75.

13. Gilbert Meilaender, "Honoring the *Bios* in Lutheran Bioethics," *Dialog* 43.2 (Summer 2004): 123.

14. Celia E. Deane-Drummond, *The Ethics of Nature* (Oxford: Blackwell, 2004), 129.

15. Celia E. Deane-Drummond, *The Ethics of Nature*, 127–28. See also Celia Deane-Drummond, *Brave New World? Theology, Ethics, and the Human Genome* (London: T. & T. Clark, 2004).

16. Jeremy Rifkin, *Algeny* (New York: Viking, 1983), 252.

17. Stanley Shostak, *Becoming Immortal* (Albany: SUNY Press, 2002), 43.

18. Shostak, *Becoming Immortal*, 2.

19. Michael D. West, *The Immortal Cell: One Scientist's Quest to Solve the Mystery of Human Aging* (New York: Random House/Doubleday, 2003), 212.

20. West, *Immortal Cell*, 226, West's italics.

21. West, *Immortal Cell*, 227.

22. Philip Hefner, *The Human Factor* (Minneapolis, MN: Fortress Press, 2000).

23. Congregation for the Doctrine of the Faith, "Introduction," in *Instruction on Respect for Human Life in Its Origins and on the Dignity of Procreation (Donum Vitae)* (February 22, 1987), Aeta Apsotolicae Sedis 1988.

CHAPTER SEVEN

~

The Future Wholeness Framework

Hippocrates, the nephew of the ancient Greek Asclepius and the man who unknowingly gave his name to the "Hippocratic Oath," is remembered for having said: "benefit and do no harm."[1] What is frequently remembered is that our doctor should "above all, do no harm" to us. This translates into the bioethical word, *non-maleficence*. Yet, a close look will reveal that Hippocrates opened the statement with a positive admonition: Work for the "benefit" of the patient. This translates into the bioethical term, *beneficence*. Beneficence and its link to a future of wholeness is the topic of this chapter.

Future wholeness is the third ethical framework for looking at the stem cell debate. Within this general frame, several approaches can be discerned. A common one focuses on medical benefits.[2] For that reason, in previous works such as *The Stem Cell Debate* by Ted Peters, the term "medical benefits" was used.[3] However, this might not say enough. What guides and energizes ethicists in this framework is a vision of what humanity can be and should be. It is a vision of human wholeness, of human well-being in body, mind, spirit, and in community as well. Physical health contributes significantly to this vision, to be sure; yet what we think of as "medical" is but one ingredient in the recipe for human fulfillment.

A vision of the future that includes physical healing for those who suffer is certainly part of the framework. And, it may be enough for some. We, the authors of this book, place ourselves within this framework; yet we ask for much more. We ask for abundant life, the kind of life Jesus in the Gospel of John claimed he would bring. Theologians who envision a better future place improved health within a more comprehensive context of the abundant life, of the divine promise of healing not only of the body but also the soul and the community. Thus, we distinguish here *medical benefits* from *future whole-ness*, identifying such benefits as a crucial dimension of human abundance. The future wholeness framework is concerned with bringing about a future that is different from the past. The pursuit of medical benefits is a significant component of actualizing that hoped-for future.

An integral connection between the science of medicine and a theo-logical vision is this: The future can be different from the past. Whereas protectionists—both embryo protectionists and human protectionists—ground their ethics in present biological nature or in assumptions about human origins, the orientation of those working from within the third framework is toward a transformed future. As the word transformed im-plies, the future wholeness framework focuses not on past forms and condi-tions, but on the future and the possibilities that it might contain.

As is true of the other two frameworks, one can be situated within the fu-ture wholeness framework and argue either *for* or *against* embryonic stem cell research. Those who argue against supporting stem cell research from within this framework contend that there is too much hype about stem cells, that the science simply may not deliver what is touted. Raising expectations and then dashing them is immoral. Yes, people who now suffer from diseases and traumas just *might* benefit from this medical science; yet, it is cruel to make promises that might not be kept. Thus, a concern for emotional well-being as part of future wholeness might lead some to oppose stem cell research for fear that it will raise expectations only to disappoint many. For example, William Hurlbut, a practicing physician and member of the President's Council on Bioethics, raises the caution that people morally opposed to the destruction of embryos might not trust the foundations of medicine if medicine draws on embryonic stem cell work.[4] In this view, future wholeness might be jeopar-dized by the undermining of trust in medicine itself, and this is a reason to oppose human embryonic stem cell research.

Another argument made by those who oppose stem cell research from within this framework is a concern for costs within a more comprehensive concern for economic justice. Because the science of regenerative medicine is so expensive, and because the medical benefits may be limited to only those segments of society who can afford it, many of the world's poor may never see such benefits. Would it not be better to take the money now invested in stem cell research and use it to fight poverty? Would we not be better stewards of our current wealth to use our resources for raising the standard of living for the poor so they can take advantage of existing medical services? If future wholeness includes the entire world and a particular concern for the poor, then attention to future wholeness would oppose any new medical technologies that appear to further disadvantage the poor or to be available only to the rich. These are the kinds of arguments regarding medical benefits that are raised against stem cell research from within the future wholeness framework.

The Moral Urgency of Scientific Research

On the other side, those who support stem cell research from within this framework argue on the basis of potential: This science has the potential for relieving suffering and enhancing human flourishing. The usual argument here is a straight argument about medical benefits: This opportunity to do good for the human race ought to be pursued. To retard or block it would be immoral.

Enthusiastic support for regenerative medicine from within this framework can be found within the scientific establishment and among patient advocacy groups such as SCAN. Both secular and religious ethicists can operate from within the future wholeness framework. We find many Jewish thinkers here, arguing that the pursuit of medical science is itself a religious responsibility. The "religious responsibility" position can be found among Jewish moralists who rely on the concept of *Tikkun Olam*—the responsibility to join God in repairing and transforming a broken world. Similarly, many Christian ethicists appeal to the concept of *agape*—understood as the responsibility to love one's neighbor. In sum, the future wholeness framework can provide a home for secular, Jewish, and Christian ethics.

Indeed, some in the future wholeness camp argue strenuously that it is immoral for us to slow down the speed of research on regenerative medicine. Every year that goes by without developing effective therapies for cancer or heart disease or Alzheimer's means another year of needless suffering and death. The countless individuals who will continue to be victims of these genetically related diseases can claim that their blood is on the hands of the embryo protectionists who were able to use public policy to shut down life-saving research.

Eric Juengst and Michael Fossil raise an argument such as this. "If ethicists or the public would restrict the uses of embryonic stem cells, then they must then bear responsibility for those patients they have chosen not to try to save by this means. Currently, patients die regularly because transplantable organs are unavailable. There is no moral culpability in this: physicians are powerless. If stem cell research can provide the power to address this need, however, the claims of those patients become compelling."[5]

When it comes to the justice concern, supporters of stem cell research generally agree that universal access to medical benefits ought to be a moral priority. Stem cell research ought not to become one more division between the world's affluent and the world's poor. We must avoid contributing to a new division between the GenRich and the GenPoor. The GenRich would be those with the financial means to take advantage of the new genetic therapies, regenerative medicine included. The term "GenRich" was coined to refer to wealthy families who could afford to enhance the genomes of their future children. Princeton's Lee Silver envisions an exacerbated division between the GenRich among future advantaged classes with enhanced genomes. He contrasts them to the "Naturals" who, due to lack of financial resources, could benefit from the advances of genetic science but are unable to avail themselves of those benefits.[6] Today's economic inequality could become expanded into tomorrow's genetic inequality. This would retard any movement toward wholeness, when wholeness is communally understood. Access to regenerative medicine, like access to genetic enhancement, signals an issue in economic and genetic justice. Human wholeness includes communal wholeness; and our community is global in scope. Thus justice is understood to be entailed by a concern for wholeness.

Medical Benefits Plus . . .

Although the three of us who write this volume come from different branches of Protestant Christianity, we are in agreement about building on an understanding of future wholeness. As we have suggested, our argument from within this framework differs a bit from a strictly medical benefits position. Our position grows from two roots: striving for human flourishing and the vision of a better future. Both include human health and well-being, but we understand flourishing and a better future in theological terms. While those who stress medical benefits may depend simply on the principle of beneficence, our approach to future wholeness draws not just on beneficence as an ethical principle but also on a sense that we are responding, as religious ethicists, to God's will and God's call.

As religious ethicists, we affirm that God intends "abundance" or "fullness" of life for all (John 3:16).[7] Fullness of life includes health. Thus, we begin by focusing on the revolutionary therapeutic potential represented by stem cell research. Along with our Jewish colleagues (see chapter 14 on Jewish and Muslim perspectives), we believe that there is a mandate to heal and to relieve suffering. Stem cells hold out hope of developing therapies for persons suffering from cancer, spinal injury, heart disease, macular degeneration, diabetes, Parkinson's, Alzheimer's, and countless other diseases that devastate many in our population. If science in the form of regenerative medical research can give expression to our compassion for those who suffer and can serve human well-being and flourishing, then it ought to do so. To do so is to anticipate the abundance that God intends.

To support science in this enterprise is therefore an act of *stewardship*. From the creation stories, we note God's command to human beings to be good stewards of the things of the earth. In a Christian view, everything belongs to God. It is given to us, or loaned to us, for our betterment, health, and flourishing. But it must always be used with attention to God's designs and God's purposes. Working toward human flourishing is a form of good stewardship of the gifts we are given by God. It supports a God-given potential for the benefit of this world.

Indeed, we believe we humans are called to be *created cocreators*.[8] We are not responsible for our being here. God is. God has created us from nothing and given us both our existence and our mandate. That mandate is to engage in creative activity. We are not God, most certainly. Yet, we have been given the talent for creative transformation; and to be good stewards of this God-given talent requires concentrated energy on making this world better if not making ourselves better.

We would be bad stewards to simply accept the world given to us, to live within its inherited limits (as the human protectionists might have us do). Rather, we have been created with the capacity to change the world, to make it a better place: more just, more loving, and more peaceful. God's creation is ongoing, and we are part of that ongoing creation.

Further, God makes promises. Some of the most powerful passages in the Hebrew scriptures (commonly called by Christians the Old Testament) have to do with the covenant that God makes with the people. God's people are promised a future that looks nothing like the past. The ancient Israelites took those promises sufficiently seriously to leave the comfort of Egypt in search of the "promised land" that would be flowing with milk and honey—symbolizing abundance. As Christians, our ethical demands are not all encapsulated in the original creation but must also attend to God's promise of a new world. Christianity is noted for seeing life not as a circle, but in more linear fashion: The future will not simply return us to a state of the past, such as the garden of Eden, but will move us forward toward God's desired ends.

This means for us that ethics must be future-oriented. To the creation stories of Genesis we must add the vision of wholeness given in the book of Revelation: No one will be excluded, there will be enough food and drink for all, the city streets will be glittering with jewels, war will cease, and there will be no more pain and suffering. Beginning with a vision of a redeemed creation with a future wholeness, present suffering becomes unacceptable. The biomedical sciences become a means by which one strives to realize elements of that future vision during the present era.[9]

Thus, the future orientation of our brand of ethics fits with the future orientation of medical research. The promise of stem cell therapies still lies in the future. We recognize that it is an uncertain future. The benefit of this research is contingent, still undetermined. Many skeptics believe the promises made for regenerative medicine will never be realized, and that the hype leads us toward unrealistic expectations. We would join with Leon Kass (see chapter 13 on Kass) and others who caution against the hype that is often associated with stem cells in the popular press. But in the place of *hype* we would put *hope*—genuine theological hope in the future. The science of regenerative medicine provides hope based on a vision of a future with improved human health and well-being.

This is another way of saying that we believe what matters most in human life cannot be derived simply from a theology of past creation. The promise of redemption tells us that our future is not restrictively determined by our past. The fact that human life has been riddled with genetically inherited disease and other forms of suffering and affliction does not have to predetermine our future as humans. We can envision with the last book of the Bible, Revelation, a future life with God in which "mourning and crying and pain will be no more" (Revelation 21:4).

Where Kass along with others in the human protection framework stress adherence to a human nature taken from our past that sets limits, we stress God's promise of abundance and of a future in which God's will might be done and all might live in peace, justice, and felicity. This is the second root, the eschatological and transformatory root, of our beneficence position within the future wholeness framework.

Is This a Utilitarian Ethic?

Some critics of the appeal to future benefits suggest that such an appeal renders the argument consequentialist or utilitarian. Loyola University ethicist Kevin O'Rourke presents the argument in a gentle but damning way: "Many scientists, ethicists, and health-care advocates maintain that because of the great good that might be achieved, the use of embryonic stem cells in research and therapy should be accepted as an ethical procedure. But in simple language, this is an attempt to state that the end justifies the means."[10]

What is being said here? In a thoroughgoing utilitarian perspective, the good that might be done by stem cell therapies could be seen as sufficient to "trump" any and all harms done along the way. To put it another way, we would sacrifice babies for the benefit of grown ups. The ends would justify any and all means. The good to be done by developing stem cell therapies would justify destruction of embryonic life. This is the position O'Rourke seems to be combating.

We argue to the contrary that our position is not utilitarian. We do not advocate sacrificing babies for grown ups. We hold with O'Rourke and others that not all means are acceptable to reach a desired end. Our view is more akin to the classical teleology of Roman Catholicism: The "right" thing to do is determined by our ultimate destiny, which is union with God. Some things would so thwart that destiny that they might never be right to do. Thus, even in the midst of efforts to do good, there can be constraints.

Non-maleficence is one such constraint. When Hippocrates said, "benefit, and do no harm," the "do no harm" clause becomes a constraint that works with the mandate to benefit. Of course, this puts on us the burden of indicating when non-maleficence puts constraints on the good that we seek and the ways that we seek it.

This means, then, that those who would support stem cell research by arguing from within the future wholeness framework need to address the question of the moral status of the blastocyst. Is destruction of the blastocyst a "harm" that should put constraints on the good of stem cell research? All too frequently, the embryo protection and future wholeness frameworks grind together like unlubricated gears. In contrast to the embryo protection and human protection frameworks, the future wholeness framework is characterized by emphasis applied to the bioethical principle of beneficence—that is, it frames the moral debate in terms of the good pursued, not the harms avoided. No compromise on non-maleficence is entailed here; but beneficence provides the orientation.

Future Wholeness and the Ethic of Beneficence

As we have said, the focus within this framework is the promise of medical research to develop therapies to help persons suffering from cancer, spinal injury, heart disease, diabetes, Parkinson's, Alzheimer's, and countless other diseases and disorders. It is the suffering of this world that demands ethical response. Compassion—suffering with others—yields a desire to remove the cause of the suffering, improve the situation, and create hope. The future wholeness framework begins by focusing on the revolutionary therapeutic opportunity represented by stem cell research. "Science in service" is the point of departure here. If science in the form of medical research can give expression to our compassion for those who suffer, and if science can serve human well-being and flourishing, then it ought to do so. To support science in this enterprise is an act of stewardship on the part of our society, an act of actualizing a God-given potential for the benefit of this world.

But what of the constraints of non-maleficence? What of the status of the embryo, then, and the possibility that all this good is achieved at the cost of doing evil? A reverse version of better-safe-than-sorry position appears among some future wholeness advocates. Given the lack of resolution about the status of the embryo, and the uncertainty as to whether it is truly "harmed" in stem cell research, moralists claim that we are not "better safe than sorry" but "better sorry than safe." We must sometimes take risks in order to do good. Where it is not clear that there is a harm involved, we can—and possibly must—move forward. Opportunity awaits for helping suffering people who could benefit from stem cell therapies. In the face of the uncertainty concerning the moral status of the embryo, those concerned about speeding up the arrival of medical benefits elect to pursue research in spite of their uncertainty. Doing nothing—or worse, shutting down stem cell research—passively violates the principle of non-maleficence as it pertains to those now suffering who could eventually benefit.

For embryo protectionists, non-maleficence always trumps beneficence, and the assumption is that the embryo is harmed in stem cell research; therefore the research is wrong. For the future wholeness moralist, there is no trumping. Rather, one must discern the strength of each ethical principle—non-maleficence and beneficence—and weigh them given particular circumstances. This method is informed by the work of W. D. Ross, who proposed that ethical principles are "prima facie" but not absolute.[11] Principles can conflict, and when they do, one must discern which makes the stronger claim in the circumstances. In circumstances where great harm can be proven, non-maleficence will be overriding. But in the absence of proven and substantial harm, beneficence will be overriding.

We who support regenerative medicine from within the future wholeness framework must factor in uncertainty. What we have are hopes and dreams that regenerative medical research will lead to breakthroughs. Yet, we must admit that this is not guaranteed. The promises of stem cell research may turn out to be excessive, unrealizable. If this turns out to be the case, then we will need to be realistic and accept it. In the meantime, we are willing to absorb the uncertainty and proceed.

Because of the uncertainty factor, a reverse better-safe-than-sorry position is insufficient to persuade some embryo protection ethicists. Lisa Sowle Cahill acknowledges the beneficence concern, while she parries her opponents by reducing their position to an alleged utilitarian one. "Research on embryonic stem cells thus presents bioethics with a classic moral dilemma: is it ever right to cause some evil to achieve a greater good? Does the end justify the means?"[12] Cahill seems reluctant to acknowledge that beneficence-oriented ethicists also believe that persons are ends, not merely means. Quite specifically, those who now live and suffer are persons; they are the ethical end of the beneficence commitment. In addition, as the reader will see later in this chapter, we do not hold that deriving stem cells from preimplantation embryos constitutes violence or sacrifice of human persons. In sum, we do not trade non-maleficence for beneficence. We plead innocent on the charge of utilitarianism.

Variants of Cahill's argument appear elsewhere. The Council on Bioethics pits President George W. Bush's principled ethics against beneficence. Acknowledging that the White House, following previous federal law, was disallowing federal funding for certain kinds of research, the Kass council wrote, "If one accepts the premise that the decision was grounded also in a discernable . . . moral aim, one cannot show that the policy is wrong merely by pointing to the potential benefits of stem cell research."[13] The promise of medical benefits fails to trump other fundamental moral principles for Kass, apparently.

Gilbert Meilaender, who also served on the U.S. President's Council on Bioethics, concedes just slightly more to beneficence while still holding firm to embryo protection: "One who looks on life this way need not, of course, suppose that beneficence is unimportant or that relief of suffering is of little consequence. Weighty as such values are, however, they have no automatic moral trump."[14] For Meilaender, non-maleficence applied to early embryos trumps beneficence for living persons who suffer from diseases or traumas. We, in contrast, deny that one trumps the other. Rather, we assert that when all the cards are laid on the table the beneficence cards will score high.

Battling the Charges of Consequentialism and Utilitarianism

Some critics of the appeal to beneficence raise the accusation that this position is reducible to risk-benefit analysis[15] or to consequentialist ethics.[16] We already addressed this earlier when denying accusations that our position is utilitarian. We stated unequivocally that the beneficence position does not say crudely that the end justifies the means.

The beneficence position locates itself within the future wholeness framework. Advocating pursuit of medical research is a stand-alone argument, operating solely within the framework of spelling out the potential for improving human health and well-being. It does not require as a premise the sacrifice of one human good for another. It certainly does not incorporate baby killing in laboratories.

We could sympathize with this criticism only if the logic of beneficence consisted of sacrificing the confirmed dignity of the early embryo on behalf of the medical benefit of other patients—that is, risking loss of one life in order to benefit another. This criticism misses its target, however, because it confuses two frameworks, future wholeness and embryo protection. Should embryo protectionists prove the contention that the blastocyst requires protective dignity, then beneficence would apply to the blastocyst; it would require us to direct medical benefits to the early embryo.

Therefore, we would like to concede an important point. The claims of the embryo protectionists regarding the dignity of the preimplantation embryo deserve attention. Those who would support stem cell research by arguing from within the medical benefits framework need to address the question of the moral status of the blastocyst. We who are authoring this book feel obligated to respond to the challenge placed before us by Vatican moral theologians. Although we operate primarily within the future wholeness framework, we will need to shift frameworks to speak directly to the questions regarding the divine impartation of a human soul, the basis for affirming human dignity, and its application or nonapplication to the preimplantation embryo.

The Moral Status of the Blastocyst

If in fact disaggregation of the blastocyst constitutes the destruction of human persons, then the argument of embryo protectionists carries significant moral weight. We feel it is necessary, therefore, to attempt to articulate a position that takes embryo protection seriously. This is complicated by the fact that the three of us do not completely agree; our own understanding of the moral status of the *in vivo* fetus differs to some extent.

As a beginning point, nonetheless, we observe that the position known as *developmentalism* attempts to hold together a concern for protecting the embryo with a commitment to foster research on medical benefits. The "bio/moral developmental position," as advocated by secular bioethicists such as Arthur Caplan, holds that the moral status of the developing embryo changes with the embryo's biology. For those who hold this position, biological capacities such as the capacity to feel or the capacity for self-consciousness represent significant ethical thresholds. Peter Singer, for example, argues that the capacity for sentience is a reasonable criterion of moral status.[17] An entity lacking sentience, then, would not be deserving of the same protections as an entity that has sentience. Under this kind of reasoning, the blastocyst would deserve few protections, as it lacks sentience, has no capacity for autonomy, and so on. It is permissible to destroy a blastocyst, when it would not be permissible to destroy an embryo later in development.

The most widely held variant of the developmentalist position is the so-called 14-day Rule. Several morally relevant biological changes occur between the embryo's 12th and 14th day of development. At 12 to 14 days the *in vivo* embryo adheres to the mother's uterine wall. The primitive streak, which marks the location of the future backbone, appears, and the central nervous system first begins to develop. At this point, twinning can no longer occur. All three of these characteristics may be morally relevant: the rudiments of the nervous system necessary for sentience, the adherence to the uterine wall necessary for any future development, and the shutting off of the possibility of twinning. Prior to 14 days, the embryo cannot (strictly speaking) be treated as a morally protectable individual, since twinning might occur. Prior to 14 days, it is not sentient. And *ex vivo* or without adhering to the uterine wall, it has no potential for future development as a person.

Taking seriously these morally relevant characteristics, then, one can say that the *ex vivo* zygote and early blastocyst do not constitute an individual human being who would possess an immortal soul in the creationist sense of ensoulment and who would therefore have inviolable dignity. We believe that the 14-day Rule provides the most adequate ethical evaluation of early embryology to date. We affirm that what we find in the petri dish up to 14 days is a gift of nature that could prompt a giant leap forward in human health and well-being. And we commend this analysis to the public policy debate. In a later chapter we provide further elaboration and justification for this position.

Global Justice: Medical Benefits for All

Most important for us is the recognition that "wholeness" is an all-encompassing term. Medical benefits are part of wholeness, but only part. Humans are moral, spiritual, and social beings as well as physical beings. Our wholeness requires all these dimensions working together for the common good. Thus, future wholeness understandably incorporates a discussion of global justice, usually in the form of economic justice. For those who appeal to beneficence, widespread inexpensive access to the benefits of this exotic form of medical therapy becomes an implied mandate. In the guidelines first drafted by the Ethics Advisory Board of the Geron Corporation, we included a statement that appeared in the *Hastings Center Report*: "it is morally paramount that research development include attention to the global distribution of and access to these therapeutic interventions."[18]

Not everyone is in favor of this vision for justice. Those who embrace aggressive capitalism and a profitable future in biotechnology find the concern for global justice to be a form of "treason" against evolutionary survival of the strong over the weak. "Reconciling the profitability of a company engaged in expensive research with the egalitarian goal of giving their products to the world's savages free of charge is more than an ethically and financially challenging task—it is impossible. . . . The doctrines of altruism and social justice—in which the lives and minds of the best men are shackled in servitude to the world's losers—would have made impossible the construction of the first mud hut, let alone industrial civilization. Yet, academics are offering it, in the name of the latest ethical theory."[19] Yes, the authors of this book are offering just such an ethic, an ethic that leads us to serve the weak and infirm through medical research. However, we do not appeal to the "latest ethical theory." We appeal to ancient foundations such as the wisdom of Hippocrates and Jesus.

Conclusion

In the meantime, it is crucial that ethical frameworks not be confused. Clarity is needed about how arguments are structured and how commitments are oriented. Those who support stem cell research from within the future wholeness framework—including ourselves—should pause to respond to the claims made from within the embryo protection framework. Conversely, the embryo protectionists should recognize the kind of argument their opponents raise. Commitment to future wholeness does not entail sacrifice of early human beings on behalf of existing ones. Nor does it rely on a crass means-end subordination of sacred life for scientific ends. Arguments developed within different frameworks entail different priorities and different suppositions. In order to bring these arguments into conversation, whether to refute one another or to complement one another, we need to be clear about how frameworks work.

Thus far we have analyzed three ethical frameworks within which moral positions are currently being taken in the debate over the permissibility or nonpermissibility of embryonic stem cell research. In each framework, we note an implicit if not explicit commitment to protecting dignity. Whether it is the dignity of the embryo, the dignity of DNA, or the dignity of potential beneficiaries of medical research, all three frameworks presume that someone or something should be treated as an end and not utilized as a mere means.[20]

This shared commitment to dignity demonstrates that all three positions should be considered *ethical* positions. None advocates something unethical or immoral. All want what is good and wholesome. That they find themselves in sometimes bitter opposition to one another is an accident of history, the inability to reconcile their respective interpretations of the scientific facts and prospects. Curiously, ethically minded people can become combatants in moral wars and cultural wars, flinging invectives like spears and firing judgments like mortar attacks. Each combatant believes he or she is right. A crucial question is where sacredness lies: in the cell, in our received "nature," or in the future promised to us by God.

Notes

1. Hippocrates, *Epidemics I:xi*, in W. H. S. Jones, *Hippocrates with an English Translation*, vol. 1 (Cambridge, MA: Harvard University Press, 1959), 165.

2. For examples of this framework see Robert P. Lanza, et al., "The Ethical Validity of Using Nuclear Transfer in Human Transplantation," JAMA 284.24 (2000): 3175–79, along with Ted Peters and Gaymon Bennett, "A Plea for Beneficence," in *God and the Embryo: Religious Voices on Stem Cells and Cloning*, ed. Brent Waters and Ronald Cole-Turner (Washington, DC: Georgetown University Press, 2003) 117–30.

3. Ted Peters, *The Stem Cell Debate* (Minneapolis, MN: Fortress Press, 2007), chapter 5.

4. This comment was made in a presentation at "The Dilemma of the Stem Cell: Research, Future, and Ethical Challenges"—a conference sponsored by the Islamic Organization of Medical Sciences in cooperation with the World Health Organization, Cairo, Egypt, November 3, 2007.

5. Eric Juengst and Michael Fossel, "The Ethics of Embryonic Stem Cells—Now and Forever, Cells without End," JAMA 284.24 (2000): 3180–84.

6. Lee Silver, *Remaking Eden: How Genetic Engineering and Cloning Will Transform the American Family* (New York: Avon, 1998).

7. The passage is often translated with the phrase "eternal life." However, the translation can also be "abundant" life, with a focus on the quality of life here on earth rather than on a future life.

8. The phrase, "created cocreator," is one of the gifts to theology offered by Philip Hefner in *The Human Factor* (Minneapolis, MN: Fortress Press, 1993).

9. See articles coauthored by Ted Peters and Gaymon Bennett, Jr., "Cloning in the Whitehouse," *Dialog* 41.3 (Fall 2002): 241–44; "Defining Human Life: Cloning,

Embryos, and the Origins of Dignity," *Beyond Determinism and Reductionism: Genetic Science and the Person*, ed. Mark L. Y. Chan and Roland Chia (Adelaide, Australia: ATF Press, 2003), 56–73; "A Plea for Beneficence: Reframing the Embryo Debate," *God and the Embryo*, 111–30.

10. Kevin D. O'Rourke, O.P., "Stem Cell Research: Prospects and Problems," *National Catholic Bioethics Quarterly* 4:2 (Summer 2004) 293.

11. W. D. Ross, *The Right and The Good* (Indianapolis, IN: Hackett Publishing Company, Inc., 1988), chapter 2.

12. Lisa Sowle Cahill, "Stem Cells: A Bioethical Balancing Act," *America* (March 26, 2001), www.americapress.org/articles/cahill-stem.htm.

13. *Monitoring Stem Cell Research*, A Report of the President's Council on Bioethics (2004), www.bioethics.gov, 26.

14. Gilbert Meilaender, "The Point of a Ban," *Hastings Center Report* 3 (January–February, 2001): 15.

15. See Julie Clague, "Beyond Beneficence: The Emergence of Genomorality and the Common Good," in Celia Deane-Drummond, *Brave New World? Theology, Ethics, and the Human Genome* (London: T. & T. Clark, 2004), 189–224.

16. "Consequentialism is the moral theory that actions are right or wrong according to the consequences they produce, rather than any intrinsic features they may have, such as truthfulness or fidelity." James F. Childress, "Consequentialism," in *The Westminster Dictionary of Christian Ethics*, ed. James F. Childress and John Macquarrie (Louisville, KY: Westminster/John Knox Press, 1967, 1986), 122.

17. Peter Singer, *Practical Ethics* (Cambridge, UK: Cambridge University Press, 1979).

18. Karen Lebacqz, Michael Mendiola, Ted Peters, Ernlee W. D. Young, and Laurie Zoloth, "Research with Human Embryonic Stem Cells: Ethical Considerations," *Hastings Center Report* 29:2 (March–April 1999): 31–36.

19. Alex Epstein, "Biotech vs. Bioethics," *The Intellectual Activist* 17:7 (July 2003): 18.

20. "Act in such a way that you always treat humanity, whether in your own person or in the person of any other, never simply as a means, but always at the same time as an end." Immanuel Kant, *Groundwork of the Metaphysic of Morals*, tr. by H. J. Paton (New York: Harper, 1948) 96.

CHAPTER EIGHT

~

Ethical Smoke and Mirrors in Washington

President George W. Bush gave his first public policy address regarding stem cells on August 9, 2001. The president supported existing restrictions on embryo research that denied federal funding for research that would entail the derivation of new human embryonic stem (hES) cell lines; but he did permit government funding for continued research on existing lines. His advisors thought the number of viable lines would be 78. Researchers in the know estimated that only a dozen or at the most 20 were characterized and reliable. By 2004 the National Institutes of Health (NIH) admitted that at most 23 lines could be considered viable.[1] Among scientists critical of the funding restrictions, these became known as the "presidential lines."

The 2001 address signaled the growing importance of stem cell research as a political issue. But has that political importance led to significant developments in understanding or to helpful policy? It has certainly served to catalyze important discussions, but to date this has not translated into better policy. As noted in the previous chapter, President Bush established the President's Council on Bioethics (PCBE), which was directed until 2005 by University of Chicago professor Leon Kass. We described some of the work of the PCBE in a previous chapter, especially the work on cloning and its indirect implications for stem cell research. The council subsequently studied stem cells directly, and the results have been at best confusing, at worst discouraging.

In this chapter we will briefly review federal policy precedents regarding the use of embryos in scientific research. Then we will turn to the president and the PCBE to track and interpret some key if controversial outcomes of their work. We examine policies on federal funding policy to show how federal leaders take ethical positions within the embryo protection and nature protection frameworks.

Funding for Embryo Research?

Most medical research in the United States is funded by the NIH and the Department of Health and Human Services. In 1996, an amendment that became known as the Dickey Amendment was added to the budget bills that fund both these agencies. That amendment specified that no funds could be used for the creation of human embryos for research purposes or for research in which embryos were destroyed or knowingly subjected to the risk of death.[2] The Dickey Amendment has been attached to the funding bill for these agencies every year since 1996.

Hence, federal policy prevented use of tax monies for creation or destruction of embryos in research prior to the isolation of hES cells in 1998. In 1999, however, the General Counsel of the Department of Health and Human Services interpreted the Dickey Amendment to permit federal funding for stem cell research using lines developed in the private sector. This is the general background for the current policies.

In his public policy address on August 9, 2001, President Bush declared that embryonic stem cell research "is at the leading edge of a series of moral hazards."[3] He linked it to cloning and to the "hatcheries" of Aldous Huxley's Brave New World. Thus, he proclaimed his belief that public policy in this arena "must proceed with great care." His conclusion was that federal funds could be used to support research on stem cell lines that had already been derived, but that no taxpayer funding would "sanction or encourage further destruction of human embryos."[4] Further, federal funds could be used for this research only if the stem cell lines had been derived in a certain manner.

With scientific wheels stuck in Washington mud, segments of society have taken matters into their own hands. The state of California passed Proposition 71 during the fall of 2004, approving a $3 billion bond measure to support stem cell research. The California Institute for Regenerative Medicine (CIRM) can now sidestep federal funding restrictions. Other states have followed suit, and stem cell research proceeds apace in Connecticut, Massachusetts, California, and a number of other states.

In 2007 Washington legislators also tried to free up federal funds for embryonic stem cell research with the Stem Cell Research Enhancement Act (HR.3 and S.5). The Senate version included funding for "alternative" means for deriving stem cells, including adult stem cells. Both the House and Senate versions would support human embryonic stem cell research, even if nonembryonic sources would also be supported. "Science is a gift of God to all of us," said Senator Nancy Pelosi (D-CA). "And science has taken us to a place that is Biblical in its power to cure, and that is the embryonic stem cell research."[5] Pelosi spoke in favor of stem cell research from within the future wholeness framework, but the president has taken a stand from within the embryo protection framework. President Bush exerted his veto power to block similar legislation in 2006, so in 2007 Capitol Hill looked for an unattainable two-thirds majority to override the veto threat. As of this writing, federal funds remain hooked to the protection of the early embryo from being used in laboratory research.

There is a telling issue here. Protection of human subjects in research is generally handled by mandating that all such research must receive approval from an Institutional Review Board (IRB). The IRB is charged with ensuring the safety and welfare of human subjects and protecting their rights of self-determination through requirements for "informed consent" and right of refusal. Should research on the blastocyst warrant IRB review? To say "yes" would suggest that the blastocyst is a human subject. To say "no" would establish acceptance of the fact that the early embryo is not a human subject. To date, U.S. federal policy does not require that IRBs treat early *ex vivo* embryos as human subjects. Where research is reviewed by an IRB, it is in order to protect the woman's right of consent in donating eggs, for instance. Stem cell researchers would like to keep it this way. The National Academy of Sciences guidelines (which are followed by most stem cell research organizations) assign to an Embryonic Stem Cell Research Oversight (ESCRO or SCRO) committee, not to an IRB, the task of overseeing the procurement and use of hES cells. This suggests some ambiguity within federal guidelines in spite of the president's strong policy statements.

How the Council Was Established

President Bush's opposition to embryonic stem cell research draws moral conclusions from ethical arguments within both the embryo protection framework and the human protection framework. Expressing his belief that "life is a sacred gift from our Creator," Bush pledged to "foster and encourage respect for life in America and throughout the world." First note the point of departure for Bush's moral reasoning: concern for the value of human life. Second, note that he applies this value to the early embryo. Third, note that for Bush human life is in a position of vulnerability (vulnerable to devaluation) relative to stem cell research. This presumption of vulnerability suggests that bioethics ought to take a defensive posture—it should defend human life against devaluation. Bush worried that stem cell research would encourage a cultural devaluation of life. Thus, we see reflected here both a focus on the embryo and a need to protect it, and also an unspoken assumption that something valuable about the "human" was being threatened.

Bush's policy address raised, but left unanswered, two key questions. First, what is it about the nature of human life that its value is threatened by advancing biotechnology such as stem cell research? Second, what is it about advancing biotechnology that it poses a threat to the value of human life? These two questions, implicit in the logic of Bush's response to stem cell research, would infuse the mandate laid upon the President's Council.

President Bush concluded his policy address by announcing his intention to form a Council on Bioethics, headed by Professor Leon Kass of the University of Chicago. The Council, as Bush's Executive Order would state, was commissioned to "advise the President on bioethical issues that may emerge as a consequence of advances in biomedical science and technology." This advisory role was to consist of five clearly defined components. These five components have served to orient the work of the Council. Here they are:

1. To undertake fundamental inquiry into the human and moral significance in developments in biomedical and behavioral science and technology.
2. To explore specific ethical and policy questions related to these developments.
3. To provide a forum for a national discussion of bioethical issues.
4. To facilitate a greater understanding of bioethical issues.
5. To explore possibilities for useful international collaboration on bioethical issues.

President Bush immediately asked the PCBE to tackle cloning and stem cells as two of the developments that had the most import for human and moral significance. The cloning report of 2002 was its first. The overall purpose of the Council was to provide a forum for the nation to discuss and evaluate the moral issues involved and to keep the nation appraised of new developments.

Stem Cells in the U.S. President's Council on Bioethics

The Council published its report on stem cells in January 2004. However, as we noted in the previous chapter, it had already indirectly forbidden stem cell research in its 2002 cloning report

In the 2004 report, the Council reviewed the science of stem cells, including developments since 2001. It reminded its readers of what was then current federal law and policy, summarized very briefly above. Particularly important there, of course, is the Dickey Amendment that prohibits federal funding for research that involves the destruction of embryos. Finally, it summarized the ethical debates, particularly those focusing on the moral standing of embryos.

The PCBE noted that President Bush's policy walked a fine line between prohibiting stem cell research altogether and permitting it to go forward, as the Clinton administration had previously done. Bush's policy would allow some research to go forward, but only on existing lines and on those lines only if they had been derived under certain conditions. The policy was an adroit attempt to please both sides of the debate—to give some support for stem cell research while keeping the federal government from being accused of any complicity in the destruction of embryos.

Thus, the PCBE declares that the question of *federal funding* for stem cell research is the central issue. Noting the significant differences in opinion about the basic issue of moral standing of the developing blastocyst,[6] the PCBE argues that its primary question is not this fundamental moral question. Rather, it focuses on the question of *whether federal funds should be used* to support this research: "the question at issue is not whether research using embryos should be allowed, but rather whether it should be financed with the federal taxpayer's dollars."[7] The Council reiterates this point several times: "it is important to remember that the issue in question is public funding, not permissibility."[8] Or, "the only issue in the present debate is one of federal funding."[9]

It is strange that in the midst of a public debate raging over whether stem cell research is or is not ethical, the Council would side-step that issue and declare that the central moral issue is federal funding. As the Council itself notes, funding has symbolic significance: "the decision to fund an activity is more than an offer of resources. It is also a declaration of official national support and endorsement."[10] The Council is correct to note this. But precisely for this reason, neither the president nor the Council can pretend that a failure to fund has no impact on fundamental moral questions. If funding implies support and endorsement, failure to fund implies lack of support and withdrawal of endorsement. Indeed, the failure to fund research on new stem cell lines has had, in the opinion of many scientists, a devastating impact on stem cell research. While the PCBE suggests that work supported by private funds can proceed without restriction,[11] in fact private funds for stem cell research dried up following the president's policy.

Here is why. Nearly all sophisticated research needs to take advantage of university scientists and university laboratories. Privately funded projects normally contract with universities. But now federal policy forbids universities that accept federal money to permit privately funded stem cell research to take place in the same building that houses any federally funded research. Because scientists working on federal grants work already in the most advanced facilities of any university, the result is that virtually all university-owned laboratory equipment is off limits to privately funded research projects. This has brought stem cell research in the United States, even privately funded research, to a near stand-still. Scientists are now looking for the next breakthroughs to take place in South Korea, Israel, China, Singapore, and the United Kingdom. California hopes to use some of the Proposition 71 money to build buildings for state-supported research that do not include facilities for federally supported research.

The PCBE also argues that, for those stem cell lines approved for federal funding, there is no limit on the amount of funds to be invested.[12] This is technically true, but it is also true that embryonic stem cell research has not received the same level of funding as has adult stem cell research, for example. Thus, despite the smoke and mirrors, a decision about funding *is* a statement about fundamental ethics, and both the president and the Council must be held accountable for stifling stem cell research. In the effort to protect the embryo and the sense of being "human," the concerns raised in the future wholeness framework have been either ignored or undervalued. The stifling of research is a significant ethical issue that must enter the debate with more prominence.

Sleight of Hand in Washington

Further, it is quite clear that the PCBE engages in a sleight of hand when it declares that the ethical issue is funding rather than the basic issue of whether or not embryonic stem cell research is morally licit. Surely the public debate has focused *not* on the question of using federal funds for this research but precisely on the question of whether the research is morally right or wrong. The president and the Council worked hard to avoid offending anyone's moral convictions, an impossible task given that the public debate is as divided and contentious as is the issue of moral standing of the early embryo. The president and the Council thought they could avoid offending those who believe that at any stage of development, a human embryo should not be destroyed. They failed, however, to attend to other, equally strong, moral convictions.

The PCBE notes that the relief of pain and suffering—what we have called the future wholeness framework with its stress on "beneficence"—is a value broadly (perhaps even universally) shared.[13] Aiding the sick is not only a good deed; it is, in the view of many, a decisive ethical obligation. We would agree with the PCBE that the duty to find cures for disease is not an unqualified or absolute duty—that is, we acknowledge that other ethical demands may on occasion take precedence. There are limits to any duty and places where even strong obligations come into conflict and must be prioritized. But the PCBE draws a false conclusion when it moves from this recognition of competing obligations to the implication that only those who fail to give the embryo moral status would support stem cell research. They have jumped to a conclusion for which they do not provide evidence.

Ironically, in jumping to this conclusion, the PCBE once again claims that the moral status of the embryo is the fundamental question. If this is in fact the case, then how can the PCBE itself side-step this ethical issue? By reducing the debate on stem cells to a funding issue the PCBE attempts to dodge the moral controversy. So what *do* the Council members say about the moral issues?

They suggest that there are two approaches to the moral status of nascent human life. One approach takes *continuity* to be central. Human life is always "in development." We pass through different stages all our lives. These stages may look different, but there is an underlying sameness and continuity. Hence, only the very beginning—sometimes called the "moment of conception"—serves as a reasonable boundary line for according moral status to a human being. Once conception has occurred and a new zygote is among us, from that point on it has moral status and must be protected. While it may change during development, there is continuity as well as change, and moral status is based on this continuity.

The other approach takes *discontinuity* seriously; it finds morally relevant differences between different stages of development. While biology itself does not yield ethical answers, some biological developments correlate to features that are morally important. For instance, as noted in the previous chapter, the appearance of a primitive streak at 12 to 14 days after fertilization is morally significant, because it is the stage after which division of the embryo into two or more identical twins can no longer occur. Prior to this time, the developing zygote might divide into two or more individuals. Since moral standing generally requires individuality, the determination that there is indeed an individual matters morally.[14]

Others suggest that the brain or nervous system is an especially important feature of human life and hence that an embryo lacking the primitive streak (the foundation for the nervous system) does not yet have an essential aspect of human character. In short, those who take discontinuity seriously argue that "there are developmental differences, and these differences matter" for moral purposes.[15] The PCBE notes that those who take such a view may nonetheless declare that an early embryo or blastocyst, while not having *full* moral status, is deserving of respect.[16]

The PCBE thus displays with some accuracy typical arguments on both sides of the debate about the moral standing of human embryos. Indeed, it reviews not only these fundamental arguments but others as well—for example, arguments for using "nonviable" embryos that would not be able to live anyway, or for creating nonviable "embryo-like" entities that are not in fact fertilized embryos. The Council tries to avoid taking sides in the contentious debate, and turns instead (as we have seen) to the question of funding as a way out.

Taking a Position in the Ethical Debate

But is this good enough? We believe that it is not possible to avoid taking a clear ethical position in this debate. If indeed, as the PCBE itself argues, funding implies support or lack of support for the research, then a position on public funding *is* an ethical position. The Council effectively takes an ethical position against stem cell research. The Council favors positions within the embryo protection and nature protection frameworks; and it fails to give priority to beneficence or future wholeness.

Finally, what if we take seriously the question about funding? Let us suppose for a moment that the issue really is the question of public funding for this research. Here, the Council does not challenge the president's policy. It appears to support it, though it never says so directly. That policy presumes that federal funds should not be used for purposes that violate the strongly held moral convictions of a selected portion of the population. Whether that portion of the population is as small as the president alone or as large as the Roman Catholic and Southern Baptist populations in the United States, the central idea is that use of federal funds may not violate deeply held moral convictions

Such a presumption clearly cannot be sustained. Use of federal funds— particularly in a congressional system that allows filibustering, lobbying, and attaching of "amendments" to appropriations—already violates some of the desires and convictions of almost every citizen's. Many Americans strongly opposed going to war in Iraq (as many opposed U.S. policies in El Salvador or in Vietnam). Such opposition is nearly always based on strong moral convictions. In any democracy tax dollars will always be used to support actions that some constituents do not approve and did not support with votes. That is the nature of democracy. It is disingenuous, then, for the PCBE to hide behind the question of public funding for research, or to suggest that because there are differences of opinion regarding the ethics of stem cell research, federal funds should not be used to support that research.

As we turn from this account of moral commitments hidden beneath the smoke screen of funding restrictions, we will clear the smoke to reveal the more profound philosophical deliberations over ethics that have taken place within the PCBE.

Notes

1. Cynthia Fox, *Cell of Cells: The Global Race to Capture and Control the Stem Cell* (New York: W. W. Norton, 2007), 182.

2. President's Council on Bioethics, *Monitoring Stem Cell Research* (Washington, DC, January 2004), 25. At the time of preparation of this report, the Council included two members whose terms were subsequently not renewed: Elizabeth H. Blackburn, professor of biochemistry and biophysics at the University of California, San Francisco; and William F. May, at that time visiting professor in the Department of Religious Studies at the University of Virginia. The nonrenewal of their terms on the Council was very controversial, and many believe that the presence of these two members may have been important to the balance that the Council brought to its presentation of differing views on the moral status of the embryo.

3. Fox, *Cell of Cells*, 185.

4. Fox, *Cell of Cells*, 186.

5. 110th Congress, 1st session; *Congressional Record* 153.91 (June 7, 2007), H6138.

6. Throughout its discussion, the Council uses the term "embryo" rather than "blastocyst." We discuss the differences in terminology throughout this book.

7. President's Council on Bioethics, *Monitoring Stem Cell Research*, 37.

8. Fox, *Cell of Cells*, 40.

9. Fox, *Cell of Cells*, 62.

10. President's Council, *Monitoring Stem Cell Research*, 37.

11. President's Council, *Monitoring Stem Cell Research*, 46.

12. President's Council, *Monitoring Stem Cell Research*, 44.

13. President's Council, *Monitoring Stem Cell Research*, 58.

14. President's Council, *Monitoring Stem Cell Research*, 79.

15. President's Council, *Monitoring Stem Cell Research*, 82.

16. President's Council, *Monitoring Stem Cell Research*, 83.

~

CHAPTER NINE

The Hidden Theology behind
the International Debates

"It's not just a matter of faith, it's a matter of science," Senator Bill Frist told the U.S. Senate on July 29, 2005. Really? Or, is it the reverse? What appears to be a matter of science—or, more accurately, government regulation of science—appears to us to be an unacknowledged matter of faith. The senator gave the okay for stem cell research using discarded existing embryos, but he forbade creating new embryos for research. Why? The answer is not a scientific answer. It is a theological answer, even when only semiarticulated. "I am pro-life," added the senator. "I believe human life begins at conception."

Like traffic police trying to direct scientific research, governmental study commissions the world over are blowing the stop whistle on reproductive cloning and turning the yellow caution light on stem cell research. Specifically, public policy formulators ask: Should experiments that might lead to reproductive human cloning be permitted? Should research on human embryonic stem cells be allowed to go forward? Questions such as these are being posed in secular venues, but responses to them carry religious substance. In fact, the public policy debate depends upon a behind-the-scenes theological debate. That debate usually centers on the origin of morally protectable human personhood, or dignity; and it is being applied to the early embryo slated for dismantling at the blastocyst stage in order to create pluripotent stem cell lines.

Despite our previous analysis discerning three competing moral frameworks—the embryo protection framework, the human protection framework, and the future wholeness framework—it remains the case that in policy debates, only the first framework is operative. These policy debates cover over a significant fact: at the heart of the debate over the protection of the embryo lies a theological debate over the nature of the soul.

As governing bodies try to map the road toward ethical research within the embryo protection framework, they are continually blocked by the question of when morally protectable human life begins. They are blocked by the theological question of ensoulment. The problem of identifying the moment when inviolable human dignity and moral protectability are established is tacitly the problem of whether or when ensoulment occurs. The theological doctrine of ensoulment assumes that the presence of the soul implies presence of a person; and the presence of a person requires an ethics of nonmaleficence. When applied to the debate over research on human embryonic stem cells, the early embryo is declared a person with dignity who deserves protection and, thereby, destruction of early embryos in order to pursue stem cell research is rendered morally illicit.

In what follows, we identify the issues as they appear in the public policy debate by referring to three important international discussions, the United Nations discussion of cloning from 2002 to 2005, the Singapore Bioethics Advisory Committee (BAC) recommendations of June 2002, and the U.S. President's Council on Bioethics report to the White House of July 2002. What we find in all of these is an easy rejection of research on reproductive cloning and a profound moral struggle over the permissibility of research on human embryonic stem cells. We will then show how the latter struggle over stem cells reflects a behind-the-scenes theological debate wherein questions of the human soul, human person, human dignity, and moral protectability arise. We will suggest that the existing set of questions poses a problem of scope and clarity, because it ignores other relevant theological considerations. Specifically, what is left out of the theological debate that we believe to be important are questions regarding the role of *relationalilty* and *eschatology* in thinking about human dignity. We will explain what we mean by these terms as we proceed.

The track we would like to follow here leads from the public policy debate over scientific research to the theological debate that underlies it. In this chapter, we will see how the embryo protection framework defines the parameters for much of the international debate. (Later, in the final chapters of this book, we will turn in the direction of constructive theological anthropology; we will attempt to provide a better understanding of the human soul and human dignity.) Audrey Chapman of the AAAS in Washington poses the challenge we would like to take up here: "The prospect of human cloning reveals the need to rethink and redevelop our theological anthropology to be consonant with the contemporary scientific understandings and issues."[1]

We do not expect to resolve the public policy debate as long as that debate focuses solely on asking just when in embryo development morally protectable dignity can be established. Rather, we plan to use the public policy debate over stem cell research as a point of departure for exploring just what we mean by human nature and human dignity. We will argue that focus on the time of ensoulment in the embryo—even if it could supply a resolution to the public policy debate—falls short of embracing the full scope of theological resources for understanding human personhood.

The Public Policy Debate in the United Nations

In the fall of 2001 France and Germany pressed a joint initiative in the United Nations General Assembly to draw up an international convention against cloning human beings. These two countries contended that a prohibition would be needed against both reproductive cloning plus cloning for research and therapeutic purposes. Under its Resolutions 56/93 of December 12, 2001, cosponsored by 50 nations, the Ad Hoc Committee on an International Convention against the Reproductive Cloning of Human Beings was established. This committee met from February 25 to March 1, 2002. Chaired by Peter Tomka of Slovakia, its report established overall agreement to prohibit human reproductive cloning (a moratorium, not a permanent ban), but no consensus could be established regarding research and therapeutic cloning. "There was general agreement that the reproductive cloning of human beings was a troubling and unethical development in biotechnology, and that it should be prohibited." Then the report went on to ground its objection on a perceived threat to human dignity. "It raised moral, religious, ethical and scientific concerns and had far-reaching implications for human dignity."[2]

The debate over cloning broke out again in 2003, 2004, and 2005. To include a ban on therapeutic cloning along with one on reproductive cloning would avoid falling down a moral slippery slope, it was thought. If the technology for therapeutic cloning could be developed, then the capacity for reproduction with this technique would become available. How might one guarantee that unscrupulous scientists would not simply take the step from one to the other? Under these pressures, on August 8, 2005, the UN voted to "prohibit all forms of human cloning inasmuch as they are incompatible with human dignity and the protection of human life."[3] Note that "all" forms of human cloning are to be prohibited, therapeutic as well as reproductive. By implication, since "therapeutic" cloning is nuclear transfer for research purposes in stem cell research, much human embryonic stem (hES) cell research would also be banned.

What is going on here? Within which ethical framework is such a discussion taking place? It appears that in this international political arena the debate is over the moral status of the human embryo. "I think the whole vote keyed on the status of the embryo," said Arthur Caplan, who made a presentation to a UN committee in November 2003. "It was a battle about the metaphysical and moral standing of the embryo."[4]

As we look at the more thorough studies pursued in Singapore and the United States, the same confident rejection of reproductive cloning will be accompanied by ambivalence over research and therapy, especially regarding hES cells. Protection of the human embryo from unscrupulous scientists has become the tacit agenda of the international debate.

The Public Policy Debate in Singapore

Here is the image of Singapore that one journalist, Cynthia Fox, presents. "[In c]ountries whose governments were not beholden to a powerful religion (China and Singapore) or whose religion did not hold the moment of conception as sacred (Israel and some Muslim nations), the door was tantalizingly open" for stem cell researchers to immigrate.[5] Singapore, she continues, "has decided to be unusually laissez-faire in the critical area of hES cells."[6] By implication, strong religions lead to restrictions on science; weak religions open the door. Does religious laxity provide an open door for stem cell researchers to avoid pesky ethics? Not in the case of Singapore, to our observation. This metropolis and country may be secular, but behind its secular policies we can discern hidden religious concerns and a good deal more ethical integrity than is popularly thought.

It is certainly the case that the government of Singapore wants to increase biotechnology investment in its country. Investment now could bring big returns in the future. Late in 2001—following President George Bush's public statement on August 9, 2001—the government of Singapore commissioned its BAC to investigate what is involved in stem cell research and to review carefully the concerns of the many religious traditions represented among its diverse populace. After hearing from the diverse religious communities, the BAC formulated its own set of regulations, which were published in 2002. Since this time, a flow of stem cell companies and researchers has streamed toward this island country just south of Malaysia and north of Indonesia. Could we say that a Whore of Ethical Babylon is enticing those who lust for a morality-free-zone in which to pursue stem cell research? No. Here's the story.

The Singapore BAC solicited theological input from the religious groups represented in this most pluralistic of societies: Hindus, Buddhists, Muslims, Orthodox Christians, Protestant Christians, and Roman Catholics. Of these groups, only the Protestants and Catholics put together teams of theologians and ethicists to think through the issues raised by genetic science in order to provide recommendations for public policy.[7] Leaders in the other religious traditions were appreciative that they would be consulted by the BAC, but they admitted they simply were not prepared to deal with such sophisticated scientific issues. When the BAC turned to make its own decisions, it emphasized that Singapore is a pluralistic society and that no single religious position could be permitted to dominate. What resulted in the BAC regulations could safely be called "secular," though religious concerns had informed its deliberations.

The Singapore BAC filed its June 2002 report after six months of study.[8] The first task was easy to dispense with, namely, setting policy for reproductive cloning. Recommendation 7 of the BAC report said: "There should be a complete ban on the implantation of a human embryo created by the application of cloning technology into a womb, or any treatment of such a human embryo intended to result in its development into a viable infant." What seems clear is that Singapore, along with the rest of the world, offers no support for procreative cloning. What is less clear is whether this lack of support is due primarily to moral disapproval or simply to the lack of a vision of how a financial profit could be made from developing reproductive cloning. With low economic interest, very little resistance to moral proscriptions seems to hold sway. Be that as it may, the cloning debate is really not much of a debate.[9] Consensus has been relatively easy to attain in Singapore.

Easy agreement was not to be found on the matter of stem cell research, however. Here controversy prevails. The controversy reverts to the embryo protection framework and centers on the derivation of hES from the blastocyst. The BAC approved the use of stem cells derived from adult tissues and also from cadaveric fetal tissues (human embryonic germ or hEG cells). With a tone of careful monitoring, the BAC invoked beneficence as a moral guide, that is, the potential benefit to human health as a moral factor contributing to the approval of research. Yet, recommendation 3 stated: "Research involving the derivation and use of ES cells is permissible only where there is strong scientific merit in, and potential medical benefit from, such research." Then it proceeded to set a sequence of preference of sources for hES cell derivation in recommendation 4: "Where permitted, ES cells should be drawn from sources in the following order: (1) existing ES cell lines, originating from ES cells derived from embryos less than 14 days old; and (2) surplus human embryos created for fertility treatment less than 14 days old." This prioritized sequence of options reflects an underlying ambiguity over the moral status of the blastocyst. The proposed sequence reflects priority given to what is most moral and in descending order to what becomes increasingly questionable.

After the options of using existing cell lines or activating a surplus embryo, might there be a third option, namely, the creation of fresh embryos for the purpose of research? Yes, but only after the first two options would be tried and found inadequate. Recommendation 5 put it this way: "The creation of human embryos specifically for research can be justified only where (1) there is strong merit in, and potential medical benefit from, such research; (2) no acceptable alternative exists; and (3) on a highly selective, case-by-case basis, with specific approval from the proposed statutory body." By the end of 2006, no scientist had applied for permission to produce fresh embryos, so this provision has yet to be tried.

The ambivalence reflected here is between giving a green light to proceed with the research, on the one hand, and protecting the early embryo, on the other hand. In competition are two ethical principles: *beneficence* or the motivation to pursue medical benefits and *non-maleficence* or the proscription against doing harm. In this case non-maleficence is applied to the early embryo. The gingerly delineation of preferences for sources from which to derive stem cells demonstrates the BAC's respect for the blastocyst; yet, in this case, the level of respect does not trump the anticipated value of what progress in medical research might yield. The BAC approach is by no means a crass utilitarianism; rather, it delicately seeks to honor the embryo protection position while delivering permission to proceed with therapeutic research that may include nonreproductive cloning.

The boudoir to which Singapore is wooing the world's scientists and biotech companies is Biopolis. The Biopolis complex consists of nine buildings (seven built, with two more under construction) and provides laboratory space for companies and universities. It provides an army of more than 500 scientists in its Agency for Science, Technology and Research, or A*Star. Alan Coleman, former member of the team that cloned Dolly, took over as chief scientist of ES Cell International (ESI) in 2002; and now ESI is attempting to induce stem cells to turn into insulin-producing islet cells to combat diabetes. Pharmaceutical companies such as Eli Lilly, Novartis, and

Glaxo-SmithKline have established research centers at Biopolis. Its annual spending for research by private sector investors tripled from 2001 to 2004, from $88 million to $238 million. In parallel, manufacturing of biomedical products tripled from $6 billion to $18 billion.

"We should strive to become a global thought leader in science and technology just as the founding political leadership made Singapore an international thought leader in public government," writes Edison Liu, who left the U.S. National Cancer Institute in 2001 to become executive director of the Genome Institute of Singapore.[10]

One might ask: Just why are genetic researchers as well as pharmaceutical researchers flocking to Singapore? One might surmise that it is due to *laissez faire* ethics. It appears that American stem cell researchers find Washington's policies too restrictive, too limiting, perhaps suffocating. Does Singapore appear to be a safe haven for scientists wanting to avoid ethical supervision?

Now, we need to look more carefully. What we see here is that the ethical discussions in Singapore are quite similar to those in other parts of the world. When it comes to deriving hES cells, priority is given to existing cell lines; the 14-day Rule is honored; and government permission for creating new research embryos is required.

Research scientists at Biopolis find government supervision in Singapore to be thorough and detailed, yet not cumbersome or obstructionist. Because the Singapore authorities have virtually eliminated corruption and because they follow the guidelines set down by the BAC, laboratory scientists actually find the firm governmental structure supportive. They gladly fill out the paperwork that ensures they meet guidelines. By no means can one accurately describe Singapore as morally lax.

Why, then, are genetic researchers flocking to Singapore? The brain flow is clearly due to wooing by the Singapore government, the establishment of Biopolis as the scientist's dream laboratory, and enormous financial incentives. The Singapore government is practical. Money talks loudest. Propping up firm ethical policies that are informed by religious communities provides Singapore with an image of credibility, and moral credibility in the long run is thought to be profitable. In the meantime, scientists are enjoying the combination of financial incentives and ethical supervision.

The BAC, headed by medical doctor Lim Pin, continues to be active, consulting with religious leaders and overseeing research. The BAC interacts with the public through its website, www.bioethics-singapore.org/ and its members, including Supreme Court Justice Richard Magnus, would be chagrined to think the world's image of this island nation is that it is lax on ethics.

Finally, Protestant and Roman Catholic spokespersons are less than fully happy with what is happening. The embryo protectionists among these two Christian groups appreciate that their voices have been heard, but they would like to see even tighter protections for the early embryo. They feel dismissed when the BAC formulates its policy by appealing to "pluralism." By appealing to pluralism, every religious position becomes likened to a personal opinion. No individual religious opinion is allowed to become public policy. BAC regulations are, thereby, strictly "secular."

The Public Policy Debate in Washington

In the United States the questions are the same, even if the answers are different. In the previous chapter we reported that the U.S. President's Council on Bioethics (PCBE), first chaired by University of Chicago professor Leon Kass, sent to the White House a July 2002 report titled, "Human Cloning and Human Dignity: An Ethical Inquiry."[11] As with the Singapore BAC, the U.S. PCBE recommendation here was for a total ban on *cloning-to-produce-children*. Such a ban is not remarkable; it reflects general worldwide opinion.

The Council took up the matter of human embryonic stem cell research under the category of *cloning-for-biomedical-research*. Although the PCBE was unanimous in banning reproductive cloning, on cloning for research the vote was split, ten to seven, in favor of a four-year moratorium. This split vote reflects the same ambiguity we find in Singapore, a debate between beneficence and non-maleficence, a debate between medical benefits and embryo protection.

The U.S. PCBE split first ten to seven, with the majority opposing stem cell research. Within the minority of seven, two subpositions developed, one to approve stem cell research without cautions and one to approve with cautions. What is important for us here is to observe what considerations entered into delineating the cautions. The cautions appear in what the report calls "position one," which strongly advocates that research proceed, albeit with strict federal regulation. The list of concerns reflects ambiguity. Acknowledged here are four ethical difficulties. First, the PCBE refers to something called *intermediate moral status*. The intermediate moral status is applied to the early embryo at the blastocyst stage. "We believe there are sound moral reasons for not regarding the embryo in its earliest stages as the moral equivalent of a human person. We believe the embryo has a developing and intermediate moral worth that commands our special respect, but that it is morally permissible to use early-stage cloned human embryos in important research under strict regulation."

The second concern is called *deliberate creation for use*. As we saw in the Singapore case, a debate is taking place within the scientific community over the derivation question: Should we limit research to discarded frozen embryos or should we deliberately create fresh embryos that will be taken apart when retrieving hES cells? Most existing stem cell lines on President Bush's list of approved lines were derived from frozen surplus embryos, although at least one University of Wisconsin line was derived from a fresh embryo. It appears that fresh embryos have research advantages over frozen ones, meaning that we can expect future pressure to produce embryos for research purposes. The council minority instructs us on language to use here. "These embryos would not be 'created for destruction', but for use in the service of life and medicine. They would be destroyed in the service of a great good, and this should not be obscured." In short, beneficence trumps non-maleficence in the minority report.

Thirdly, the PCBE is concerned about *going too far*. This minority position wants to exact a time restriction, to limit the size of the window for pursuing research. They do not want early embryos to develop too long so that they approach the fetus stage. Here the 14-day Rule enters the picture. "We approve, therefore, only of research on cloned embryos that is strictly limited to the first fourteen days of development." This fits with the Singapore BAC and the widely acknowledged primitive streak threshold. As noted in an earlier chapter, somewhere between 12 and 14 days the *in vivo* embryo adheres to the uterine wall of the woman, a primitive streak appears which will eventually become the backbone. This is a threshold because here for the first time we have an individual fetus that could become a child.[12] So, this minority position within the Council supports the 14-day Rule.

The fourth concern has to do with *other moral hazards*. Increasingly we hear voices in the public debate registering concern over the justice implications of gathering human eggs for scientific research, justice concerns that focus on the health and well-being of women who donate the eggs. The Council has heard these voices. In addition, a slippery slope fear has arisen. Some fear that if we approve cloning in stem cell research that we could slide gradually toward approving reproductive cloning as well. The Council's minority confronts these matters by saying, "We believe that concerns about the exploitation of women and about the risk that cloning-for-biomedical-research could lead to cloning-to-produce-children can be adequately addressed by appropriate rules and regulations. These concerns need not frighten us into abandoning an important avenue of research."

Of the two minority positions approving *cloning-for-biomedical-research*, "position one" alerts us to the concerns listed above. The second minority position, "position two," supports *cloning-for-biomedical-research* without any timidity. "Because we accord no special moral status to the early-stage cloned embryo and believe it should be treated essentially like all other human cells, we believe that the moral issues involved in this research are no different from those that accompany any biomedical research." What we have here is categorical support for stem cell research. This second minority position exhibits consistency—that is, because the blastocyst does not have morally protectable status it follows that research is permissible under all circumstances. The ambivalence is gone.

The two minority positions lost out to the majority. Why did the majority propose a four-year moratorium applicable to all researchers regardless of whether federal funds are involved? Here is their reason. "We believe it is morally wrong to exploit and destroy developing human life, even for good reasons, and that it is unwise to open the door to many undesirable consequences that are likely to result from this research. We find it disquieting, even somewhat ignoble, to treat what are in fact seeds of the next generation as mere raw material for satisfying the needs of our own." What the PCBE said to the president in June 2002 was said to the president a year earlier in July 2001 by Pope John Paul II during a presidential visit to the Vatican: "Experience is already showing how a tragic coarsening of consciences accompanies the assault on innocent human life in the womb, leading to accommodation and acquiescence in the face of other related evils such as euthanasia, infanticide and, most recently, proposals for the creation of embryos for research destined to be destroyed in the process."[13] What we see in the PCBE is a theological commitment being recommended to a secular government. Now, we are by no means suggesting that secular governments should not listen to theological arguments; to the contrary, we applaud this government for listening. Yet, this warrants here a review of the underlying theological debate contributing to the ambiguity in public policy.

Altered Nuclear Transfer

In 2005 the PCBE published a document entitled *White Paper: Alternative Sources of Human Pluripotent Stem Cells.*[14] In this book-length exploratory study, four proposals are offered for deriving embryonic stem cells without disaggregating embryos. Why might such a set of proposals be studied?

Quite obviously, the PCBE wants to press science into the service of embryo protection. The theologically based defense of the early embryo drives the PCBE's prompting of the scientific community to honor the moral protect-abiliy of *ex vivo* blastocysts.

One of the proposals, *Altered Nuclear Transfer* (ANT), is sponsored by PCBE member William Hurlbut. Hurlbut is a passionate defender of the embryo against laboratory destruction. He feels the moral burden of protecting the embryo while still looking for a way to create lines of pluripotent hES cells. Hurlbut earnestly desires to see stem cell derivation proceed, but not at the cost of doing harm to an already established human life.

Altered nuclear transfer, as the name implies, begins with nuclear transfer—that is, with cloning-for-biomedical-research. An enucleated egg becomes the recipient of another DNA nucleus. The key is this: Prior to the transfer of the DNA nucleus into the enucleated egg, the laboratory technician would dismantle the ability of the DNA nucleus to develop into a human being. One way to do this, for example, would be to silence the Cdx2 gene. The result would be that the nonembryo produced would be incapable of attaching to the uterine wall or developing a basic body plan. What would be lost is the global ability to develop. The nonembryo's cells would become pluripotent, but not totipotent. This nonembryo would still produce fully functional stem cells, contends Hurlbut.

Scientifically, the ANT proposal is subject to several difficulties. First, at the time when Hurlbut made his proposal, no method for nuclear transfer in human beings had been achieved. The January 2007 claim of Stemagen that it has indeed succeeded in finding a method to do nuclear transfer in humans may obviate this first difficulty, but at the time of this writing, it is too early to tell whether stem cells can easily be obtained in this way and whether they will prove to have the requisite characteristics that have made embryonic stem cells so advantageous.[15]

Second, and more troubling, the genetic crippling of the DNA nucleus might very well be passed on into the pluripotent stem cells; and they may pass the defective DNA into a future patient. Stem cell scientists, facing enormous technical challenges, desire to work with the highest quality genetic material possible. They are not likely to select to work with genetically damaged material. We also note that, at the time when Hurlbut made his proposal, other members of the Council resisted it because they saw it as an attempt to create a *defective* embryo and then destroy it for research purposes; this offended their sense of the need to protect the most vulnerable humans.

Our point in entertaining Hurlbut's ANT proposal here is to demonstrate the persistence of theological commitments marching through ethics and right into proposals for scientific research. By no means do we object to such influence of theology on science, let alone on ethics. We celebrate it. We are simply trying to contribute to truth in labeling, so to speak. And, in addition, we would like to point out that the theological debate is not over. Neither Jewish nor Islamic bioethicists would agree on a strict proscription against blastocyst disaggregation, because these traditions employ a 40-day threshold after fertilization for personhood. These religious positions would not rely upon the same theological motives for pursuing ANT or other research programs aimed at avoiding laboratory use of embryos. Be that as it may, if the research Hurlbut is sponsoring should yield success, and if others can be convinced that the entity created is indeed not an "embryo" (or a defective embryo) at all, then we could have embryo protection right along with hES cells.

The Theological Nondebate over Cloning

As we have observed, a moral consensus against reproductive cloning seems to have emerged, even when support for therapeutic cloning is sustained. It appears virtually unanimous for both Muslims and Christians.

A ban on reproductive cloning is called for among both Muslim and Christian commentators. As we report in another chapter, one Islamic spokesperson suggests it is Satanic. "Unanimity has now emerged among Muslim scholars of different legal rites that whereas in Islamic tradition therapeutic uses of cloning and any research to further that goal will receive the endorsement of the major legal schools, the idea of human cloning has been viewed negatively and almost, to use the language of the Mufti of Egypt, 'Satanic.'"[16] Bioethicist Stephen Garrard Post affirms this position with slightly understated rhetoric. "Christians side with the deep wisdom of the teachings of Jesus, manifest in the thoughtful respect for the laws of nature that reflect the word of God. Christians simply cannot and must not underestimate the threat of human cloning to unravel what is both *naturally* and *eternally* good."[17]

"Both naturally and eternally good," says Post. Does this mean the embryo protectionists and the human protectionists are allied on the cloning issue? Yes. Leon Kass speaks to what is natural, saying that cloning is "a major violation of our given nature as embodied, gendered, and engendering beings—and of the social relations built on this natural ground."[18]

But what is eternal? Calum MacKellar of the Scottish Order of Christian Unity speaks to this. Each person has "three cocreators (mother, father and God)," and genetic twinning that denies the role played by one of these cocreators "dissociates creation from the communion of triune other-ness of God and the two parents."[19] Reproductive cloning can be rejected categorically, so it seems, as a violation of nature and as a denial of divine communion. What we see here is an alliance between the embryo protection framework and antagonists working from the naturalist or anti-brave new world framework.

The Overt Theological Debate over Stem Cells

Perhaps we can set aside further analysis of reproductive cloning for the mo-ment, given the widespread moral consensus. But stem cell research and its connection to "therapeutic cloning" is another matter, a much more delicate matter that is less firmly decided. The moral convictions and ambiguities driving public policy regarding scientific research come in large part from religious convictions and unresolved theological debates that underlie them. What is debated in theological circles is being taken very seriously by secular bodies, both the scientific community and governmental agencies. An ex-amination of the theological commitments at work will aid in understanding the moral sensibilities at work in public policy deliberations over the accept-ability of human embryonic stem cell research.

Unfortunately, too often Christian commentators limit themselves to only one ethical framework, the embryo protection framework. The National Council of Churches of Singapore states the logic flatly: "If indeed an embryo is regarded as human life from the point of fertilization, then it follows that one should oppose not only reproductive cloning but also so-called therapeu-tic cloning that entails the destruction of embryos."[20] If this logic obtains, reliance upon this framework alone is understandable. However, we believe that "the point of fertilization" does not in itself set the initial conditions for all further moral deliberation.

We sympathize with regulatory agencies who are trying to direct traffic coming from two directions. Moving in one direction are those who want to expand research programs on embryonic stem cells. Moving in the other direction are those, seeking to protect the early embryo . The result is a traffic light that says both "go" and "stop."

Notes

1. Audrey R. Chapman, *Unprecedented Choices: Religious Ethics at the Frontiers of Genetic Science* (Minneapolis, MN: Fortress Press, 1999), 124.

2. "Report of the Ad Hoc Committee on an International Convention against the Reproductive Cloning of Human Beings," February 25–March 1, 2002, United Nations General Assembly, Official Records, Fifty-seventh Session, Supplement No. 51 (A/57/51). Dignity concerns were voiced earlier by UNESCO: "Practices which are contrary to human dignity, such as reproductive cloning of human beings, shall not be permitted." UNESCO Thirty-first General Assembly, "Universal Declaration on the Human Genome and Human Rights," *Journal of Medicine and Philosophy* 23.3 (1998): 338.

3. UN Press Release: "General Assembly Adopts United Nations Declaration on Human Cloning by a Vote of 84-34-37," www.un.org/news/Press/docs/2005/ga10333 .doc.htm, accessed February 7, 2008.

4. Cynthia Fox, *Cell of Cells: The Global Race to Capture and Control the Stem Cell* (New York: W. W. Norton, 2007), 169.

5. Fox, *Cell of Cells*, 75.

6. Fox, *Cell of Cells*, 121.

7. See National Council of Churches of Singapore, *A Christian Response to the Life Sciences* (Singapore: Genesis, 2002).

8. Bioethics Advisory Committee of Singapore, *Ethical, Legal and Social Issues in Human Stem Cell Research, Reproductive and Therapeutic Cloning*, report submitted to the Ministerial Committee for Life Sciences, June 2002.

9. Even Ian Wilmut, who cloned Dolly the sheep, has taken a consistent stand against reproductive human cloning. "The Roslin Institute and PPL Therapeutics have made it clear that they regard the idea as ethically unacceptable." Ian Wilmut and Donald Bruce, "Dolly Mixture," in *Engineering Genesis*, ed. Donald Bruce and Ann Bruce (London: Earthscan Publications, 1998), 75. The original scientific study that initiated the cloning controversy is by Ian Wilmut et al., "Viable Offspring Derived from Fetal and Adult Mammalian Cells," *Nature* 385 (1997): 810–13.

10. Edison Liu, "Rise of a Biomedical Dragon," *Straits Times*, November 4, 2006, S9.

11. The President's Council on Bioethics, *Human Cloning and Human Dignity: An Ethical Inquiry*, report submitted to U.S. President George W. Bush, July 10, 2002. Controversy erupted within the PCBE regarding the precise voting margin for and against cloning—for—biomedical research. See "President's Bioethics Council Delivers," *Science* 297 (2002): 322–24. For a content analysis, see the review by Ted Peters and Gaymon Bennett, "Cloning in the White House," *Dialog* 41.3 (Fall 2002): 241–44.

12. Fourteen-day Rule advocates note how the appearance of the primitive streak *in vivo* marks the earliest point at which one should ascribe moral status to the human embryo, because this point marks organized development and the onset of what will become sentience for an individual. See *Report and Recommendations of the National Bioethics Advisory Commission*, vol. 1.6, 10. See the careful treatment by the National Council of Churches of Singapore, *A Christian Response to the Life Sciences* (Singapore: Genesis Books, 2002), 91–101.

13. "Pope John Paul II Addresses President Bush," www.americancatholic.org/News/StemCell.

14. U.S. President's Council on Bioethics, *White Paper: Alternative Sources of Human Pluripotent Stem Cells* (Washington, DC: President's Council on Bioethics, 2005).

15. Constance Holden and Gretchen Vogel, "A Seismic Shift for Stem Cell Research," *Science* v.319, 1 February 2008: 560–63. While there is considerable excitement in the scientific world over these new announcements, scientists emphasize that considerable testing will be needed before it is known whether these cells are truly pluripotent. In addition, there is some caution in that they may prove useful for research purposes but not for therapy, since the creation of stem cells from nuclear transfer uses a source that may have been altered by aging or toxins.

16. Abdulaziz Sachedina, "Human Clones: An Islamic View," in *The Human Cloning Debate*, ed. Glenn McGee (Berkeley, CA: Berkeley Hills Books, 1998), 240–41.

17. Stephen Garrard Post, "The Judeo-Christian Ethic Opposes Cloning," in *Cloning: For and Against*, edited by M. L. Rantala and Arthur J. Milgram (Chicago: Open Court, 1999), 158. Italics added.

18. Leon Kass, "The Wisdom of Repugnance," in *The Ethics of Human Cloning*, Leon R. Kass and James Q. Wilson (Washington, DC: American Enterprise Institute Press, 1998), 23.

19. Calum MacKellar, *Creation, Co-Creation and the Ethics of Pro-Creative Cloning*, pamphlet prepared by the Scottish Order of Christian Unity, Edinburgh, page 1.

20. National Council of Churches of Singapore, *A Christian Response to the Life Sciences*, 97.

CHAPTER TEN

∼

The Vatican's Strong Stand

Ethicists, theologians, scientists, politicians, and others from around the world work within multiple distinct ethical frameworks as we have described. At the same time, the most widely circulated and contested arguments come from the embryo protection framework. Of the positions developed within this framework, those articulated by the Vatican's moral theologians are the most sophisticated and forceful. The authors of this book find the Vatican arguments the most thorough and well thought through of any that compete in the public sphere. Although we think that the future wholeness framework should structure reflection on stem cell research, we recognize the depth and significance of the Vatican position, as well as its influence on policy and public debate. A more thorough examination of the arguments of embryo protectionists within the Vatican is warranted.

When we say "Vatican" here, we mean to designate the official *magisterium* or teaching authority of the Roman Catholic Church, represented by papal pronouncements, by the Congregation for the Doctrine of the Faith, by Church Councils, and by other official church bodies, such as the National Conference of Catholic Bishops in the United States. Because official Roman Catholic views are quite thoroughly developed and because official spokespersons have been forceful in lobbying for specific policies, the Vatican has become the unappointed voice of global Christianity. Other Christian voices are drowned out, and Vatican stances are taken to be "the" Christian view, even though many Christians—even many Roman Catholics—disagree. As a result, the way the ethical question gets formulated is everywhere influenced by the Vatican's formulation. This is neither a compliment nor a criticism, merely an observation. Because the Vatican position is often taken to be representative of all Christians, it is particularly important to analyze its views.

It is out of profound respect for the leadership of popes John Paul II and Benedict XVI that we now turn to a more thorough analysis of Vatican thinking. Even though it is our judgment that the better route to follow is to be found within the future wholeness framework, pausing to map the Vatican route through the embryo protection framework is worth the investment of attention. If we are to provide sustainable theological support for the pursuit of regenerative medical research, we feel obligated to address Vatican claims directly, to understand them empathetically, and to provide reasonable justification for following an alternative route.

Human Embryonic Stem Cells? No!

The Vatican stand on embryonic stem cell research is strong and clear: "*Is it morally licit to produce and/or use human embryos for the preparation of ES cells? The answer is negative.*" This is the question and answer as formulated in the "Declaration on the Production and the Scientific and Therapeutic Use of Human Embryonic Stem Cells." The answer is elaborated thus: "the ablation of the inner cell mass of the blastocyst, which critically and irremediably damages the human embryo, curtailing its development, is a *gravely immoral* act and consequently is *gravely illicit*."[1] Richard Doerflinger, a policy developer for the U.S. National Conference of Catholic Bishops, puts it this way: "intentional destruction of innocent human life at any stage is inherently evil, and no good consequence can mitigate that evil."[2]

When the pope, the Congregation for the Doctrine of the Faith, and moral theologians oriented toward Vatican policy weigh in on cloning and stem cells and related issues, they appeal again and again to two precedents, *Donum Vitae* (1987) and *Evangelium Vitae* (1995). A central argument of these documents is that the zygote, the egg fertilized by the sperm, is characterized by the dignity of human personhood, and must be protected accordingly. "The Church has always taught and continues to teach that the result of human procreation, from the first moment of its existence, must be guaranteed that unconditional respect which is morally due to the human being in his or her totality and unity in body and spirit: The human being is to be respected and treated as a person from the moment of conception; and therefore from that same moment his rights as a person must be recognized, among which in the first place is the inviolable right of every innocent human being to life."[3]

This respect for morally protectable dignity is derived from the Vatican's understanding of the human soul. Such respect is reflected in Pope John Paul II's 1996 elocution on evolution: "It is by virtue of the spiritual soul that the whole person possesses such a dignity even in his or her body. Pius XII stressed this essential point: If the human body takes its origin from pre-existent living matter, the spiritual soul is immediately created by God."[4] The logic at work here is foundational. It goes like this: first, God creates a new soul and imparts it to the zygote; second, the presence of the soul establishes dignity; therefore, third, dignity prevents the use of the zygote for research purposes or its destruction in order to create stem cells. In what follows, we will tease this argument in more detail.

As an aside, we note that John Paul II commits himself to the classical doctrine of creationism—the idea that God creates a new soul that begins at conception but lasts to eternity.[5] The pontiff posits a new soul and he posits it at the beginning. What is less clear, as we will show later, is just when the beginning occurs and when the impartation of the created soul can be said to be accomplished.[6]

If having an imparted soul is central, then the precise moment at which the soul is imparted to the zygote becomes crucial in evaluating the ethical status of embryological research. Official teaching argues that it is at the "moment of conception," with the creation of the zygote. Contemporary embryology, however, tells us that there is no single "moment"; conception is not a moment but a *process*. This raises some problems for establishing the precise moment of ensoulment in biological terms. *Donum Vitae* addresses this problem through the concept of potentiality. Because the zygote has a unique genetic code, the Vatican argues, it is a physical entity crying out for a spiritual soul. Because of its potential ensoulment and its potential dignity, we should protect it from destruction at the hands of the laboratory technician.

What We Appreciate

There are several things that we value in the official Roman Catholic view. First and foremost, we appreciate the effort to protect "innocent" or vulnerable human life. Although we will offer some strong criticisms below, we do not want to be misunderstood on this point. The Vatican deserves the respect it has gained as a champion of the poor, the weak, and the vulnerable. As Christians, we share a concern to protect people who cannot protect themselves.

Second, we appreciate the effort to present a carefully reasoned argument. Too often, when emotions run high, logic flies out the window. Our Roman Catholic interlocutors work hard to offer a reasoned and logical view. We share a respect for rational argument as a basis for public policy. Indeed, it is precisely because of our deep respect for the importance of careful thinking that we respectfully disagree with some of the Vatican's conclusions.

Why We Disagree

In spite of these appreciations, we do disagree with the Vatican stance, even from within the embryo protection framework. We disagree with several of the Vatican's basic suppositions, and therefore to the conclusions drawn from these suppositions. First, the Vatican assumes that the destruction of the embryo in stem cell research can be equated with elective abortion. We argue that this ignores morally relevant differences. Second, the Vatican argument depends to a large extent on specific if not peculiar assumptions regarding ensoulment. In a later chapter, we will tackle this question directly and offer a different understanding of ensoulment. Third, the Vatican assumes that personhood is established metaphysically by the early embryo's ensoulment, regardless of whether this embryo is situated within a mother's body with the potential for developing into a baby. Whereas the Vatican assumes that ensoulment and dignity apply to the *ex vivo* blastocyst in a laboratory petri dish, we believe the relationship of an *in vivo* embryo to its potential mother contributes significantly to its moral status. Whereas the Vatican depends strictly on a metaphysical individualism to establish personhood and moral protectatibility, we view human dignity more relationally, more contextually.

Now, we would like to explain what we mean. Let us describe the Vatican logic a bit more thoroughly. Taking a number of official Roman Catholic documents together, seven principal commitments can be discerned. First, the moral concern regarding stem cell research is the same as that of abortion. Second, procreation requires heterosexual intercourse and consists of the merging of two gametes, an egg from the woman and a sperm from the man, combined with the impartation by God of an immortal soul.[7] Thus, there are three partners in the generation of new human life: a man, a woman, and God. Third, God creates and imparts a soul to a unique individual, that is, to an individual with a unique genome. Fourth, the merging of sperm and egg is something *natural*, thereby making it both the natural norm and the moral norm. Fifth, because this process is natural and because the immortal soul is then present, the embryo from conception forward claims morally protectable human dignity. Sixth, dignity requires that the early embryo be treated as an end in itself and not as a means to a further end. Seventh, it is therefore morally illicit to sacrifice the life of the early embryo on behalf of some further end, no matter how noble that end might be. It is morally illicit to sacrifice the innocent life of a person in a petri dish for the purpose of developing medical therapies to benefit others. When applied to stem cell derivation, these assumptions yield a proscription against research with human embryonic stem cells.

The Ghost of Abortion

One of the ghosts that haunts the debate on deriving embryonic stem cells is the memory of the abortion debate in the 1970s, in which positions were taken and arguments waged. The Vatican is consistent in its belief that it is always morally illicit to destroy innocent human life. That view stands behind both its opposition to abortion and embryonic stem cell research. Unfortunately, while consistency is laudable, in this case the consistency of stance ignores some morally relevant differences.

The Vatican is not the only church organization to believe that a stance on stem cells must be consistent with a previous stance on abortion. "The General Assemblies of the Presbyterian Church (U.S.A.) have consistently supported women's right to choose an abortion based on conscience and religious beliefs. . . . We believe that the use of tissue derived from fetuses is morally and ethically acceptable."[8] We see here that the Presbyterians share with other churches the assumption that the focal question has to do with protecting the dignity of the zygote. Yet, Presbyterians make a commitment that is just the reverse of the Vatican. Because this church body denied moral protection to the fetus during the abortion controversy, it feels it is being consistent when approving destruction of the blastocyst for human embryonic stem cell research. Whether pro or con on embryo protection, the controversy presumes that the dignity question will be answered by what is biologically innate, by a quality inherent or absent in the status of the unborn.

It is unfortunate that the abortion ghost haunts the stem cell controversy, in our opinion. To reduce the question of stem cell derivation to ethical commitments previously established when arguing over elective abortion inappropriately excuses theologians from grappling with what is distinctive regarding the new knowledge about the early embryo being gained by contemporary laboratory study. The abortion controversy of the 1970s and the stem cell controversy three decades later should each be treated on their own merits. There are morally relevant differences.

Here are some of the salient differences. Whereas in the earlier abortion controversy the competition was between the right of the unborn child to be born over the mother's right to choose what happens to her body, no such battle of competing rights is at stake in the stem cell controversy. Whereas in the earlier abortion controversy the unborn child exists *in* the mother's womb with the potential of a healthy birth and a normal life, the extra-utero blastocysts in stem cell research have no potential for birth because they will never see the inside of a womb. Whereas in the earlier abortion controversy the existence of a fetus in the womb could be identified only after many weeks of gestation and when the fetus is considerably more developed, the early embryos used in research are restricted to the preimplantation stage, prior to the time when they would be capable of adhering to the uterine wall and prompting a pregnancy. Thus, there are differences in the parties involved, in the potentiality of the embryo, and in the developmental stage at which the status of the embryo is considered. All of these differences are ignored by those who equate abortion and stem cell research.

What the two controversies share is belief in morally protectable person-hood at conception. We argue, however, that the lines that divided the pro-life from the pro-choice sides in the earlier abortion controversy need not divide in the stem cell controversy. Despite the logic of the Presbyterians and that of the Vatican, one could in principle approve of stem cell research and still remain opposed to elective abortion. In fact, many are in this camp. At least one of the three authors of this book is in this camp.

A Genetically Unique Individual?

In addition to the ghost of abortion, the embryo protection position relies on a version of naturalism. It presumes that nature tells us what is right. It presumes that what happens in nature has a purpose, and that what is morally licit conforms to the purpose inherent in nature. Nature can be our guide, because it is guided by God.[9] If babies are born as a result of intercourse, then this must be what God has ordained. What is "natural" is what is meant to be, and anything "unnatural" is wrong. It then becomes our ethical task to see to it that these purposes inherent in nature are carried out. Efforts to thwart or sidestep or modify nature can be dubbed illicit.

On the list of divine intentions expressed in nature, according to the Vati-can, is that each human person should be unique—genetically unique—and result from the intercourse of one man and one woman. The genomes of these two combine to create a new first-time genome, and this is the moment God blesses by imparting an immortal soul. The impartation of the immortal soul is what defines an individual as an individual, as a person with moral protectability. In effect, genetic uniqueness has become the sign of metaphysical uniqueness; and this uniqueness is what establishes personhood.

What we would like to point our here is that, inadvertently, the metaphysical and ethical logic of the Vatican and its fellow travelers has tied ensoulment to genetic uniqueness. We believe this connection between soul and genetic uniqueness is unfortunate, because it fails to take into account what science can tell us about embryo development. Efforts to hinge moral status on genetic uniqueness then render theology vulnerable to charges of incredulity.

Just how did this situation arise? We see in *Donum Vitae* and elsewhere how the merging of sperm and egg is considered natural. Also, the establishment of a unique single genome is considered natural. And, further, the Vatican sees in the "natural" order a divine intention. The message nature appears to be giving us—a message that the Vatican hears as the voice of God—is that, when the genetic code of the father and the genetic code of the mother combine into a single new genome, a historically unique person is for the first time established. This apparently awesome moment seems just right for God to honor it with the impartation of a freshly created soul. A new soul for a new individual. That is the basic logic.

This logic is shared by others. John Breck, an Orthodox theologian, makes this clear. Breck reports, "The Orthodox Church has always taught that human life begins at conception, when a sperm unites with an ovum to produce a genetically unique, living being."[10] Breck assumes here a connection among three items: fertilization, genetic uniqueness, and moral protection.

Let us look at this assertion again: The church has always taught that human life begins when a sperm unites with an ovum to produce a genetically unique, living being. One might ask: Can we say the church has always taught this when the very knowledge of genetic inheritance is less than a century old? We will not ask this embarrassing historical question. Yet, we will ask: Is it scientifically reasonable to make such a claim? No, it is not. Looking at this more closely is the task of the next chapter.

Notes

1. Pontifical Academy Life, "Declaration on the Production and the Scientific and Therapeutic Use of Human Embryonic Stem Cells," Vatican City (August 2000), online at www.cin.org/does/stem-cell-research.html.

2. Richard Doerflinger, "The Policy and Politics of Embryonic Stem Cell Research," *National Catholic Bioethics Quarterly* 1.2 (Summer 2001): 143.

3. John Paul II, *Evangelium Vitae* (March 25, 1995), Acta Apostolicae Sedis 1995, 87, 401–522; cf. also Congregation for the Doctrine of the Faith, *Instruction on Respect for Human Life in Its Origins and on the Dignity of Procreation (Donum Vitae)* (February 22, 1987), Acta Apsotolicae Sedis 1988, 80, 70–102.

4. Pope John Paul II, "Evolution and the Living God," in *Science and Theology: The New Consonance*, ed. Ted Peters (Boulder, CO: Westview, 1998), 151.

5. The term *creationism* in this context should be clearly distinguished from its use in the evolution controversy. In the evolution controversy, *creationism* in its biblical and scientific forms refers to a belief held in many Protestant circles that the earth is young, less than 10,000 years old, and that all species were specifically created by an act of God. In Roman Catholic circles, *creationism* refers to the belief that God creates a new individual soul for each person being born. The two uses have no connection with one another. More on this appears in a later chapter.

6. Anne M. Clifford is critical of papal creationism. "*Creationism* fails . . . because it holds that God directly intervenes, adding the soul, at a moment of time the human reproductive cells unite (or possibly at some point shortly thereafter). . . . The creationist understanding, however, ignores the findings of geneticists that indicate that traits such as intelligence and mental aptitudes, e.g., for music and mathematics, develop within limits that are genetically inherited. Are these things distinct from the human soul? To what then does the soul refer in a creationist understanding after modern genetics has pruned so much away?" Clifford, "Biological Evolution and the Human Soul: A Theological Proposal for Generationism," in *Science and Theology: The New Consonance*, ed. Ted Peters (Boulder, CO: Westview, 1998), 169.

7. Conception here follows two philosophically discernable stages: *active conception*, wherein the sperm penetrates an ovum and establishes genetic uniqueness; and *passive conception*, wherein God imparts the spiritual soul. Some Roman Catholics will assert that both occur at the same time, at the alleged "moment of conception." What we see in the papal documents is a disjoining of the two stages.

8. "Statement on Stem Cell Research," adopted at the 213th General Assembly of the Presbyterian Church (U.S.A.) in June 2001; online at www.aaas.org/spp/dser/news/presbyterianstatement.htm.

9. "Natural law is the normative theological and metaphysical order that undergirds, makes possible, and flows into our moral logic. . . . The natural law is normative only because it is rooted in God in whom absolute being, good, and truth are identified." Steven A. Long, "Reproductive Technologies and the Natural Law," *National Catholic Bioethics Quarterly* 2.2 (Summer 2002): 221–22, 224.

10. John Breck, *The Sacred Gift of Life* (Crestwood, NY: St. Vladimir's Seminary Press, 1998), 259. For additional Orthodox contributions to the discussions of cloning and stem cells, see Demetri Demopulos, "A Parallel to the Care Given the Soul: An Orthodox View of Cloning and Related Technologies," in *Beyond Cloning: Religion and the Remaking of Humanity*, ed. Ronald Cole-Turner (Harrisburg, PA: Trinity, 2001), 124–36; and Archimandrite Makarios Griniezakis, "Bioethical Dilemmas through Patristic Thought," *Human Reproduction and Genetic Ethics: An International Journal* 8.2 (2002): 32–37.

CHAPTER ELEVEN

~

The Vatican and Embryology

We concluded our exposition of the embryo protection position and the Vatican's strong stand in the previous chapter by noting how significant, though tacit, a role is played by genetic uniqueness. Let's rehearse the logic: At conception a unique genome appears; this biological uniqueness establishes a human individual; to this individual God imparts a spiritual soul; because of this ensoulment the zygote possesses dignity; and because of this dignity it becomes illicit to disaggregate the blastocyst to harvest pluripotent stem cells. Even if we are unable to discern scientifically the moment of impartation of the soul, the unique genome renders this human individual a person who is ready to receive that soul. Because we are unable to discern scientifically the moment of ensoulment, we must then refrain from hurting the conceptus, even *ex vivo*. Rather, we must protect it as a human person in potential. This protection includes sustaining its life just as we would any other person.

What we wish to do here is ask: How might this metaphysical understanding of ensoulment fit with what we now know about the development of the early embryo? Does our knowledge of embryogenesis complement this theological position? Even though neither the Vatican nor those writing this book would ask science to take over the job of theological reflection, we still ask whether some level of consonance exists between our theological commitments and the natural world seen through scientific lenses. It is our observation that we have a misfit here. Genetic uniqueness seems to us to be a horse the theologian should not bet on.

A Genetically Unique Individual? Really?

The problem is that there is more variation, more chance, and less surety in nature than the embryo protectionist view assumes. The moment of conception may be the moment in which a unique genome is established, to be sure; but it is not the moment in which a new individual person is created. Nor is it the case that each new human person possesses a single unique genome. Nor does the possession of a unique genome necessarily correlate with the receipt of a "soul."

Let us examine more closely the presumed correlation among individual identity, soul, and unique genome. Three phenomena occurring within nature are relevant to assessing this correlation. The first is fetal wastage. Estimates of the number of naturally fertilized eggs that are flushed from the mother's body before they can adhere to the uterine wall range from 50 percent to 80 percent. Consider how many unique genomes depart from the woman's system! What is not known is whether the woman's body expels the fertilized ova or if they depart of their own accord. What is known from the theory of evolution is that nature is profligate with regard to offspring—that is, each species produces far more offspring than are needed for sustaining the species. Nature seems almost prescient that most will die and only a (small) percentage survive to reproductive age. Nature seems quite content to eliminate the vast majority of fertilized ova and retain only a few to bring to birth. If the Vatican is serious about associating a divine soul with each and every zygote, and if the woman's body by nature eliminates the majority of ensouled embryos, then theologically it would be difficult to see God's intentions as carried out by natural processes.

This precise problem has led ethicists such as Benedict Ashley, O.P., and Kevin O'Rourke, O.P., to speculate that "probably many of these imperfectly fertilized ova were never prepared for ensoulment."[1] Note what they assume. Flushed ova are "imperfect" and are not yet prepared for ensoulment. Yet they clearly have a unique genome. Thus, a unique genome alone cannot be the basis for ensoulment. Apparently, then, it takes something more than a unique genome in order for God to create a special soul; and the flushed embryos do not meet the specifications warranting ensoulment. We see this as grasping at metaphysical straws in order to avoid the inconsistencies in linking a unique genome with ensoulment and hence, with moral protectability.

Second, there is the phenomenon of twinning. In the early embryo, each cell is totipotent—that is, each cell can make not only any tissue in the body, it can also make an entire person. In the first few days, the agglomeration of cells can divide into twins, quadruplets, octuplets, or in principle even into 16 individual embryos. All of these would have the same genetic code, even if they become separate individuals. Monozygotic twins—what we call "identical" twins—are the result of such cell division. If identical triplets are born, most likely the early embryo had split into four and one of them was flushed from the mother's body at some point. Further, during these early stages, which can last up to 12 or 14 days, these divided embryos can recombine. Twins can become a single person again. It is possible that each person reading this was once a twin at an early stage of embryonic development, even though they now are individuals. All this is possible because the cells that are dividing during early embryo development are not yet differentiated but are totipotent.

The result of the twinning process, of course, is that two or more babies can be born with identical genomes. Yet each is clearly an individual. Indeed, even conjoined twins are often noted to have very distinct personalities, though they share not only a genome but a good bit of their physical bodies. In nature, one can be a unique individual without having a unique genome. The connection between genetic uniqueness and individual personhood is therefore not a scientific judgment; it is a theological overlay. This overlay has led some radical Roman Catholic ethicists to suggest that twinning is unnatural, that twins are an aberration and should be seen as freaks. To be a twin, according to this logic, is to be ontologically outside God's intention.

Such extreme interpretations represent a minority view. What they show us is the interpretive task involved in taking "nature" as a guide to God's intentions. In classical Roman Catholic moral theology, not everything that occurs in the natural world is God's intention. As such it is crucial to know how we are to move from the biologically "natural" to the realm of divine intention. There can always be a basic incongruity between what we see in nature and what is taken to be normatively "natural." We will return to this problem below.

The third phenomenon within nature that challenges the Vatican association of an individual human person with a unique genome is chimerism. A chimera is a single individual with two or more genomes. Within the mother's body, *in vivo*, frequently two or more eggs can be fertilized at the same time. If two separate fertilized eggs develop simultaneously, each with its own amniotic sac, two babies will be born at the same time. We know these as "fraternal" twins—that is, twins with different genomes.

However, something else can take place during the first few days of embryonic development. This pair of zygotes can combine to form a single embryo. If brought to term, the resulting baby is a chimera, a single person with two genetic codes. If the two fertilized ova are of the same gender, then the baby girl or baby boy may grow up, live a normal life, and never know that he or she began as fraternal twins. If, however, a male and female combine, then the resulting baby may be a hermaphrodite possessing both male and female characteristics.[2] In such a case, a genetic test is likely to reveal two genomes, one with XY chromosome and the other with XX.[3]

In other words, here is an individual with not one but two genomes. Does this person then have two souls? Or, does God create one soul, a single soul for a single person? The attempt to link individual humanity with a unique genome simply unravels at this point. What the Vatican should have done theologically, in our opinion, is identify ensoulment with the human person and not with the genome.

Ensoulment in the Vatican

Is there any room within Roman Catholic theological discussions for alternative interpretations and even flexibility on discerning the origin of personhood? Specifically, we need to ask: Must Roman Catholics be committed to ensoulment limited to the so-called moment of fertilization? Does the Vatican argument become reducible to a biological argument? If the latter is the case, might an alternative biological argument be persuasive, at least for some Catholics?

As mentioned above, the door to this discussion swings on two hinges, *Donum Vitae* of 1987 and *Evangelium Vitae* of 1995. For the sake of consistency, both of these documents must demonstrate continuity with the 1974 *Declaration on Procured Abortion*, which expressly avoids commitment to the precise moment at which God infuses an immortal soul into the mortal body. "This declaration expressly leaves aside the question of the moment when the spiritual soul is infused." The 1974 Declaration considers the possibility that fertilization might occur first; then it would be followed by a time lag before the person is infused with an immortal soul. Still, even without the infusion of a soul, moral protection is afforded to the zygote. "Supposing a later animation, there is still nothing less than a *human* life, preparing for and calling for a soul in which the nature received from parents is completed."[4] Though the zygote does not yet possess a spiritual soul, it is calling for a spiritual soul. Nature is calling for spirit. In order for the process to be completed, the combination of egg and sperm require the intervention of God in imparting that soul.

The Vatican clearly takes a creationist position; God directly imparts an individual soul to each person. What is less clear is whether the moment of soul infusion is at conception or later. The 1974 *Declaration on Procured Abortion* does not absolutely require that ensoulment take place at conception. Getting ready for ensoulment is good enough to establish moral protection. In summary, the argument stands that even prior to the divine act of ensoulment the zygote is made for personhood, and that requires protection.

Donum Vitae continues the previous line of thinking, emphasizing that to determine just when ensoulment occurs is a philosophical task, not a scientific one. It takes philosophy to discern spiritual reality. Biological science alone cannot. "Certainly, no experimental datum can be in itself sufficient to bring us to the recognition of a spiritual soul; nevertheless, the conclusions of science regarding the human embryo provide a valuable indication for discerning by the use of reason a personal presence from the first appearance of a human life: how could a living human creature not be a human person?" On the one hand, science is disqualified from providing direct knowledge relevant to the presence of a spiritual soul. On the other hand, science prompts reason that in turn prompts awareness of a "personal presence." Now, personal presence is being defined here apart from the soul. And this personal presence "demands the unconditional respect that is morally due to the human being in his bodily and spiritual totality."[5]

Donum Vitae relies upon "recent findings of human biological science" to make the point that "in the zygote . . . resulting from fertilization the biological identity of a new human individual is already constituted."[6] If we equate identity with personhood or with dignity, then moral protectability is established at the point where the genome of the man and the genome of the woman combine to create the identifiable genome of the zygote. No soul is required, according to this phase of the argument; only a new genome is needed.

In *Evangelium Vitae*, Pope John Paul II says that "modern genetic science offers clear confirmation" of what the Church has always taught about such matters, namely, "It has demonstrated that from the first instant there is established the structure or [*structuram seu*] genetic program of what this living being will be: a man [*hominem*], and indeed this individual man [*hunc hominem individuum*]." Herein lies the warrant for moral protection. "The Church has always taught and continues to teach that the result of human procreation, from the first moment of its existence, must be guaranteed that unconditional respect which is morally due to a human being in his or her totality and unity as a body and spirit."[7] Note the assumptions on which the pope's argument relies: First, he is assuming procreation requires a fertilized egg (made up from sperm and egg), not considering a clone which may in principle survive and mature with DNA from only one source; second, that

DNA programs in advance "what this living being will be"; third, that at fertilization a person as an individual and not a group is determined and identified; and, fourth, that biological uniqueness established by DNA is sufficient to warrant protection as if we were treating a person as a totality, as both "body and spirit."

Catholic ethicist Norman Ford's interpretation is instructive here. He says that "John Paul II makes it clear that the Magisterium made no decision for or against the common opinion held for centuries that ensoulment and the beginning of the person did not occur for several weeks after conception."[8] From the Middle Ages until just recently, Ford reports, it was commonly accepted in the West that the human person did not begin until several weeks after conception. Only after Pope Pius IX in 1869 declared excommunication for aborting fetuses in all stages of pregnancy did the view begin to grow that personhood is tied to conception. This means that connecting morally protectable dignity with the moment of conception is relatively recent, and that the Church has not in fact "always taught" what is now claimed.

What about the 14-Day Rule?

This makes room for further discussion. Ironically, there is wisdom in the older Church tradition that coincides with modern biology. As we now know that the zygote may become more than one individual, assigning a later time to personhood or moral protectability makes a better "fit" with modern science. The original genetic code is capable of making zero, two, four, or even eight human beings with the same genome, depending on various embryonic processes. The cells of the early embryo do not become an identifiable individual human being until after they have adhered to the uterine wall, which begins about the sixth or seventh day. At about 12 to 14 days the primitive streak appears, defining each embryo as an individual. "In short, it can be argued," says Ford, "the presence of the genetic code itself does not suffice to constitute a human individual, but that only its activation does, whereby specialized cells and membranes are produced to form and enclose an organized human individual about fourteen days after fertilization. If this argument is accepted, fertilization is not the beginning of the development *of* the human individual but the beginning of the formative process and development *into* one (or more human individuals). Ultimately this issue cannot be resolved in the first instance by appealing to the teaching of the Church, but only by reflection and critical analysis on all the relevant scientific information interpreted in the light of sound philosophical principles."[9]

Note that Ford refers to 14 days after fertilization. It has become a commonly accepted scientific convention to permit research on embryos up to 14 days after fertilization but not beyond that time. It is often called *the 14-day Rule*. (Others, such as the California Institute for Regenerative Medicine, are more conservative, limiting research to 12 days postfertilization.) The 14-day Rule appears again and again in research guidelines in the United Kingdom, the United States, Australia, Singapore, and elsewhere. Decisive here is the observation of British philosopher Mary Warnock: In the mother's body (*in vivo*) something dramatic happens around the 14th day, namely, the embryo adheres to the uterine wall and the primitive streak appears. With the primitive streak, a significant threshold is crossed. For the first time, we have implantation; an individual appears; all cells are organized into a single organism; neural development has begun; and from this point on stem cells lose their pluripotency because they begin differentiation.

What we have here is an observation of what happens *in vivo*, and research ethicists borrow this observation to set a rule applicable to research *in vitro*, outside the mother's body, *ex vivo*. Despite the significance of this threshold, however, nature does not actually draw a bright red ethical line at 14 days, nor at any other day, for that matter. Embryonic development is a continuous process.[10] Nature does not do our ethics for us. Nevertheless, one thing very important to the Vatican is decided by nature at this time, namely, whether or not we have an individual human being and, if it is implanted in the woman's body, whether it has the possibility of growing into a human person.

Roman Catholic ethicists might follow Ford's lead and take advantage of the 14-day Rule by connecting individuation with attachment to the uterine wall instead of with the initial moment of fertilization. To do so is not necessarily to approve of human embryonic stem (hES) cell research, however. Ford still holds the early embryo in respect as potential human life and recommends "banning destructive research on human embryos."[11]

Thomas Shannon makes a parallel ethical argument yet draws the opposite moral conclusion. Embryo protectionist arguments such as we have seen from the Vatican are frequently embellished by acknowledging that the blastocyst is not an individual person but rather a potential person, but that, nonetheless, this potentiality warrants treatment with full personal dignity. Shannon contends to the contrary that potency is not fact, that a person *in potential* is not an actual person just as an acorn is not an actual oak tree. The argument from potentiality is not philosophically sufficient; so Shannon can admit that he is not persuaded that the human blastocyst is an individual human person in the strong sense of the term. Shannon sees merit in the 14-day Rule as the point where individuality and moral protectability are established; so, on this basis, he permits human embryonic stem cell research.[12]

Shannon asks contemporary embryology to inform the reason employed by theologians. What embryology tells us is that what has been assumed to be a "moment of conception" is not really a moment after all. Rather than an instant in time, conception is a process. One could define conception in terms of syngamy, the 24-hour process whereby a sperm enters an ovum and establishes a diploid set of chromosomes—that is, the establishing of genetic uniqueness in the zygote. However, as we have seen, even though the zygote may possess genetic uniqueness it is not the ontological beginning of a human individual. The better way to understand conception, according to Shannon, is this: Conception is the establishment of an individual fetus who could become a person—that is, conception is a near two-week process leading to implantation. In traditional Roman Catholic theology, individuality has been a necessary, though not sufficient, condition for personhood. Biologically speaking, individuality is not assured until implantation. Therefore, implantation at 14 days rather than fertilization on the first day appears to Shannon to provide the better conclusion for the conception process.[13]

The anthropologies of Ford and Shannon have garnered critics who dispute the 14-day Rule. Lisa Sowle Cahill says "the counterargument is that as long as an embryo is a developing life within a human genetic code, it is a person despite its uncertain identity and prospects."[14] Although Cahill herself finds the attempt to locate a defining moment to establish moral protectability a near fruitless enterprise, she recognizes the role that genetic uniqueness plays in the counterargument.

Other critics try to counter the 14-day Rule by demoting twinning to something unnatural, so as to connect genetic uniqueness with individuality in God's eyes. As we mentioned above, they affirm the Vatican view that genetic uniqueness established at fertilization coincides with individual personhood and then claim that twinning is an abnormal form of gene expression. What is normal is that unique DNA belongs to a unique person, even if nature occasionally prompts natural twinning prior to 14 days. "Twinning in the human species is a reproductive abnormality since it is disadvantageous both to the mother and the offspring," write Benedict Ashley and Albert Moraczewski.[15] Note the appeal to teleology here: Twinning is abnormal because it is disadvantageous. They argue against the idea of delayed hominization until 14 days because, among other things, "twinning is a form of cloning that is not artificial but results from an embryological accident. A clone presupposes the existence of a previous unified living organism of the same species and not a mere collection of cells."[16] Because the fertilized ovum already at the first cell division is "organized" and not merely an agglomeration of totipotent cells, they argue, twinning is abnormal, and this renders the original zygote—not each twin—the status of a morally protectable person. Twins are freaks, not natural. But, we might ask: do they each have their own soul? This argument fails, in our judgment, because it leads to a *reductio ad absurdum*.

In summary, it would appear from a first reading of Vatican statements that morally protectable personhood or dignity is dependent upon infusion of the spiritual soul into the physical body. A closer look, however, will show that the Vatican position designates fertilization—fertilization understood genetically and biologically without reference to a spiritual soul—as that which establishes personhood and moral protectability. The genetically unique zygote anticipates a future infusion of soul, and this anticipation of an eventual totality inclusive of body and soul counts morally. Further, individual personhood is what is intended by our DNA, and multiple persons who share the same genotype are considered abnormal. Genetic uniqueness alone allegedly specifies what is to be considered an individual human person, and this alone allegedly establishes moral protectability. Clones, whether deliberate or accidental, would be considered abnormal, unnatural. Given these assumptions, how would the Church finally justify protecting the dignity and personhood of twins and clones, should they ever sit in the pews?

As debates within Roman Catholicism make clear, the precise connections among ensoulment, individuality, personhood, protectable dignity, and biological development are subjects of contention. If the establishment of a genetically unique zygote anticipates a future soul, we might ask: What is the importance of anticipation? What are the implications of gaining status based on the assumption of a future completion? Our own argument is future-oriented too, though differently conceived. Here is a place where we as Vatican critics might find some commonality with the Vatican. Much of Roman Catholic moral theology is "teleological"—it finds in the proper "ends" or "telos" of things their ethical status. So do we. Yet when it comes to the early embryo, the Vatican surrenders its future orientation in exchange for sole reliance on the past: moral protectability is determined by the zygote at its origin. The effective assumption, it seems, is that the telos is contained *within* the origin. The fertilized egg has only one destiny—to become a child.

Might we consider the possibility that a zygote in a laboratory could have more than one destiny? In an *in vitro* fertilization (IVF) clinic, this zygote could be placed within a woman's uterus; then, nourished by a relationship with a healthy mother, it would eventually become a child. Alternatively, this zygote could be placed in a petri dish; then, if regenerative medicine is successful, it could become life-saving tissue for a heart patient.

The Theological Implications of Parthenogenesis

Could parthenogenesis come to the moral rescue? If laboratory scientists could activate a woman's egg without introducing sperm and draw off pluripotent hES cells at the blastocyst stage, would this solve the Vatican's ethical problem? If the invocation of natural law in Roman Catholic ethics includes the necessity of a genetic component from the father, then would an embryo with only the mother's genome not count as a protectable human? Because surgical removal of hydatidaform parthenotes has already received Vatican sanction—such removals do not constitute an abortion—perhaps use of activated eggs for hES cells could escape Vatican moral rejection. Would parthenogenesis load the ethical gun with a silver bullet?

We have two precedents to bring to bear. First, some women experience partial parthenogenesis when a hydatidiform mole begins to grow. A hydatidiform cyst is an activated ovum in which the chorionic villi swell. The growing embryo is not viable. Frequently, such hydatidiform moles are removed surgically. Roman Catholic moral theologians do not categorize this surgical removal as abortion, because the embryo is incapable of full human development.

The second precedent is found in *Donum Vitae*: Human beings born without any connection with sexuality through twin fission, cloning, or parthenogenesis "are to be considered contrary to the moral law, since they are in opposition to the dignity both of human procreation and of the conjugal relation."[17] What is being said here? The activation of a woman's ovum that leads eventually to the birth of a healthy child would be immoral on the grounds that it bypasses the "conjugal" relationship. A parthenote would not have two parents. When *Donum Vitae* was written more than two decades ago, no possibility existed that such a child might be born. Today parthenogenesis is a laboratory research project, and such a possibility is no longer unrealistic. Will *Donum Vitae*'s demand for a father's sperm in each child lead eventually to the absurd conclusion that walking and talking parthenotes are less than fully human?

Turning back to stem cell research with these precedents in mind, we note that experiments on primates as well as on human material are showing the possibility of laboratory parthenogenesis. The activation of a female egg with the full complement of chromosomes (46 chromosomes in human beings) has been accomplished. Researchers are trying to bring such activated parthenotes to the blastocyst stage for derivation of stem cells.[18]

One of the scientific as well as theological questions to ask would be this: Would the cells of the activated egg be totipotent or only pluripotent? If the former, and if such an embryo when implanted in a mother's body could in principle lead to the birth of a child—a virgin birth, curiously—what then? Because no father is involved, would this child still be considered human? Would the child's genome still call out for a spiritual soul? Or, because it lacks a male genetic component, would we use natural law criteria to dub it nonhuman and nonensoulable? Or, would the Vatican switch its criterion? Would *egg activation* become the new criterion—replacing genetic novelty— for determining an embryo's protectable moral status?

We may have the opportunity to pose these questions in concrete form. A team of Russian and American researchers headed by Elena Revzova and Jeffrey Janus of Lifetime Cell Technology in Walkersville, Maryland, has created embryos by activating an unfertilized egg. The egg was activated by chemicals rather than by sperm. The team cultivated the embryos long enough to harvest stem cells. The stem cells expressed the same proteins as normal embryonic stem cells, and they proliferated for 10 months. They are capable of producing the three primary tissue types.[19] Does this mean they are pluripotent? Totipotent? If the latter, would one want to admit it?

One reflective moral theologian has registered an opinion regarding the parthenogenesis proposal: Yes, healthy parthenotes would be human persons, and, therefore, research on parthenogenesis should be halted. If we could grow an activated egg with 46 chromosomes "to the point that *stem cells* can be derived and differentiated into various body cell types—then what we have is an embryo," writes Mark S. Latkovic. "I think that this research ought to be absolutely discontinued in humans. We should strongly presume (unless somehow conclusively shown otherwise) that the parthenogenetic embryo is a human being—and not an organism closer in kind to a hyda- tidiform mole or teratoma."[20] The logic here looks like this: Even though a parthenogenetic blastocyst does not meet the conjugal standards of *Donum Vitae*, it would still be a human being and, as such, even in the laboratory its destruction should be rendered morally illicit. In sum, evidently no ethical silver bullet can be found in parthenogenesis.

Symbolically, the "virgin birth" plays no small role in many Christian minds. If the Messiah was not the result of a conjugal relationship but of a "virgin birth," it would be a most curious thing to deny to the Son of God/ Son of Man the status of human dignity! Regardless of what brings a child into the world, Roman Catholic moral theologians are most likely to dub that child human and to grant that a spiritual soul has been imparted. These arguments *ad absurdum* are offered here only to illustrate the implications and difficulties of some of the Vatican assumptions. Science is prompting new questions that will require new theological thinking. To theologians of every religious persuasion, these questions might appear a bit curious. Nevertheless, such questions ought not to be frightening or daunting.

Theological Implications of Cytoplasmic Reprogramming

Perhaps the obvious should also be said. When it comes to zygotes activated in the petri dish, we are not relying strictly upon those made the old-fashioned way, the way assumed by the Vatican. Cloned embryos would not require a man and a woman, for example; yet when activated, a cloned zygote would eventually produce a blastocyst. Although to date it seems almost established that we will not produce babies through reproductive cloning; therapeutic cloning is still being pursued by researchers. If eventually successful, what we will find in the petri dish will not have originated according to Vatican assumptions about what is natural. Yet, it might still be life-giving. This future life-giving potential is for us an ethical concern.

We can illustrate the importance of orienting an ethical view toward origins or toward future fulfillment by a quick excursus into cytoplasmic reprogramming. By cytoplasmic reprogramming we are referring to what the PCBE has called "somatic cell dedifferentiation." The theory holds that once the DNA nucleus of a differentiated cell, say a skin cell, is returned to its dedifferentiated state, then it can be reprogrammed by the cytoplasm to become pluripotent.

On the list of scientific questions is this one: How does the cytoplasm program the DNA nucleus so as to express the genes that make specific tissue? Once this is learned and technical control of gene expression is attained, then perhaps the cytoplasm in virtually any somatic cell could be reprogrammed for specific gene expression. By reprogramming the cytoplasm in a somatic cell, the need for ova would evaporate, and so would the need for cloning. We could imagine the following scenario: For each patient on site in a clinic, a skin cell could be taken and, after returning the DNA to its pluripotent state, the cytoplasm could be reprogrammed to turn on selected genes to make selected tissue. This genetically matched tissue would then be surgically implanted, and new organ tissue would grow.

We're not there yet; but we're on the way. A breakthrough at Harvard announced in 2005 shows that this path is worth following. Kevin Eggan and his colleagues fused skin cells with hES cells. The cytoplasm in the hES cells has the potential for programming development, just as ova do. The "somatic genome was reprogrammed to an embryonic state." In drawing out the implications, the Harvard team concluded that "eventually, this approach might lead to an alternative route for creating genetically tailored hES cell lines for use in the study and treatment of human disease" and "these fusion studies [might] lead to an understanding of the factors needed for reprogramming."[21] The implications are yet to be determined, but it is not unrealistic to forecast how scientific dreams can become reality.

We have been moved closer by breakthroughs announced in 2006 and 2007. The first comes from Japan. By simply splicing four genes into a mouse skin cell, Shinya Yamanaka of Kyoto University accomplished the key reprogramming task.[22] The skin cell now functions almost as the egg does in precipitating pluripotent DNA activity. Could this be done with human skin cells? Yes. The mouse experiment was followed up by a human experiment, and with success. Yamanaka and colleagues sent a retrovirus into the skin cell of a 36-year-old woman that turned on genes ordinarily active during early embryo development, genes that produce four proteins (Oct4, Sox2, eMye, and Klf4). Calling the result "induced pluripotent stem cells" or iPS cells, these scientists claim that the cells behave like pluripotent hES cells. Technically, it is gene expression rather than the cytoplasm that is reprogrammed here; yet the point is that an adult somatic cell rather than an embryonic cell becomes the source of pluripotent stem cells. A risk in this reprogramming technique comes with the fact that all of these proteins are carcinogenic. One, eMye, can precipitate tumors almost immediately, while the other three may involve a delay. James Thomson at the University of Wisconsin is running parallel experiments of his own that avoid reliance on eMye.

Also announced in 2007, Harvard scientists activated mouse zygotes after a chromosome transfer, suggesting that patient-specific stem cell lines—human stem cell lines drawn from mouse cytoplasm—might be developed through cytoplasmic reprogramming.[23] The expression of human genes would trigger mouse cytoplasm so that eventually a distinctively human line of pluripotent stem cells would result. What the Harvard experiment adds to the Kyoto reprogramming is the mouse/human factor, the chimera factor. Both methods appear to avoid the problem of destroying human embryos to obtain embryonic stem cells. Have we obviated all the arguments of the embryo protectionists? Will reprogramming bring peace to our moral conflict? An immediate moral "yes" to the Kyoto experiment came from a spokesperson for the U.S. Conference of Catholic Bishops, Richard Doerflinger, who was quoted saying that this procedure "raises no serious moral problem, because it creates embryonic like stem cells without creating, harming or destroying human lives at any stage."[24]

On the one hand, it appears that reprogramming passes the Vatican test. Yet, on the other hand, we need to take a closer look than Doerflinger does at the implications of earlier Vatican commitments. What Doerflinger does not yet see are the anthropological implications should such experiments in cytoplasmic reprogramming or gene reprogramming prove successful in human beings. Such a breakthrough would indicate that reproduction could occur by using any healthy cell in the body. Making pluripotent stem cells in this manner is right next door to making totipotent cells. This means we could make a baby from a skin cell. Reproduction would not be limited to gametes alone. Making babies the old-fashioned way might become just that, old fashioned. Given how Vatican theologians have already attributed morally protectable dignity to future human beings at the most primitive stage of genetic potential, the same logic might require duplicate application to every somatic cell in the human body when subjected to reprogramming. Embryo protectionists might have to reexamine their biological ontologies and their theories of natural law.

Commenting in *The Hastings Center Report* on the work of Yamanaka and Thomson, Insoo Hyun makes it clear that even if iPS research goes forward, so also must hES cell research. The success of iPS developments will be measured according to the hES standard. "Human iPS cell research must proceed together with human embryonic stem cell research for many important reasons,"[25] meaning that former moral objections to hES cell research will not be obviated by iPS advances. Hyun then registers concern for new complications arising for informed consent with iPS materials; and he follows this with an observation about embryo protection. "Perhaps iPS cell researchers will discover that skin cells can be driven back even further in development to a totipotent state—that is, to a single zygote-like cell capable of generating not only all three germ layers but also all the supporting extra-embryonic tissues. If this were to happen, then one could argue that any cell in a person's body has the biological potential to give rise to another complete human being. . . . Such a circumstance would be truly equivalent to human cloning in the original horticultural sense of the Greek word *klon*—that is, 'twig.'"[26] In short, iPS will keep ethicists up late at night pondering the significance of what science is teaching us.

Relevant here is that the scientific objective of reprogramming is not to provide a new means of reproduction; rather, it is to provide a genetic match—a patient specific match—for rejuvenating tissue. However, in the process of learning how to return an already differentiated somatic cell to its predifferentiated or pluripotent state will provide the knowledge of how to make a totipotent cell—that is, a cell capable of producing a baby. In principle, then, a baby could be made from virtually any healthy cell in the human body. Let us emphasize this point: A baby could be made from virtually any healthy cell in the human body. Is this likely to occur? No, because it is not on the research agenda. However, what it reveals about nature should be considered theologically relevant.

Given the premises of both pro and con contestants in the stem cell debate, might we ask if cytoplasmic reprogramming could trigger new ethical issues? One focal ethical question would have to do with the presumed special status given to the zygote and the embryonic stem cell. This question gains its energy from the thought that early stage stem cells are potential human beings. Now we must ask: If virtually any somatic cell within our body is a potential human being, what does this do to the status of the early embryo? Does it retain a special status? If the early embryo and every other somatic cell due to laboratory technology gain the same status—the same potential to become a human being or even twins—then how would the principle of dignity apply? Could we any longer distinguish the presence of dignity for a fertilized ovum while denying it to pluripotent cells or to the wide array of somatic cells? Or, we might also ask: Should we have been applying dignity to cells rather than to persons in the first place?

Perhaps we should ask whether for theological or other reasons we may have been implicitly granting special status to gametes (sperm and egg) prior to fertilized ova. In times past, using sperm and ova appeared to be nature's way; and nature's way could be lifted up into natural law theory. We note how the Vatican's theological arguments depend on fertilization and genetic uniqueness that results from the combining of sperm and egg. However, we are learning that nature does not require this. The production of Dolly the sheep through nuclear transfer has demonstrated that neither fertilization nor genetic uniqueness is required for procreation. In the event that cytoplasmic reprogramming of a somatic cell becomes capable of producing a totipotent cell—that is, a cell capable of producing a baby—then even the gametes could become unnecessary. Each healthy cell in our body would by itself hold the status of a potential embryo.

What this implies, we believe, is that totipotency in the early embryo is relativized. Its status is not obviously unique, let alone morally absolute. A cell is not sacred. Theologians need to ask whether morally protectable personhood relies upon special status attributed to biological entities such as zygotes. We ask such questions because of the apparent shifting ground for established arguments on behalf of the alleged origin of dignity. If dignity is grounded in genetic uniqueness or some special status given to some totipotent cells but not others, then how would such an ethical position avoid arbitrariness? If dignity is closely allied to only some options within nature for potential baby making, and if other options such as that of a nonfertilized cloned embryo are excluded, then does this call into question the universal scope of human dignity? Would dignity be afforded only to children coming from fertilized ova and be denied to clones or even to monozygotic twins? Would a cloned person lack dignity because she or he lacks genetic uniqueness, or because she or he lacks two parents? Would a person born from a reprogrammed somatic cell be vulnerable to denial of dignity because she or he was not derived from a fertilized zygote? To dub such possibilities as unnatural or abnormal will not suffice here, because someday persons born this way may be walking among us.

Nature's Potentials

A final item must be dealt with in our consideration of the Vatican and embryology: the question of *potentiality*. The Vatican is careful in its arguments about genomic novelty not to say that the new genome constitutes a new person. In other words, the argument from genomic novelty is not an argument that concludes that the embryo is a person with dignity. Rather, the embryo is an individual human being with the intrinsic potential to become a human person. The embryo is a potential person, and this potentiality is given and assured by the presence of a novel genome. The crucial point is that the embryo *potentiality* is given the same moral weight as *actuality*. In other words, the embryo, as a potential person, is treated with the same respect due an actual person.

Embryonic stem cell research, however, teaches us something new about potentiality and the early embryo that we did not know before. Given what rides on the argument from potentiality, this new scientific lesson is no small matter.

Let's consider again the arguments offered by the Vatican concerning the status of the embryo. The instruction *Evangelium Vitae* states that "from the time that the ovum is fertilized, a life is begun which is neither that of the father nor the mother; it is rather the life of a new human being with its own growth." The phrase "its own growth" is crucial here. It suggests that this new life has an intrinsic potential to become a person. Indeed, this potential is connected to biological nature: the genetic code. "[M]odern genetic science offers clear confirmation. It has demonstrated that from the first instant there is established the program of what this living being will be: a person, this individual person with his characteristic aspects already well determined." The argument is simple: by virtue of genetics, the fertilized ovum is programmed to become an individual person.

But is this really true? Is it the case that a genetic program alone provides the conditions for the creation of an individual person? This view certainly conforms to popular understandings of how genetics work: that genes determine what an organism will be. Research on embryonic stem cells, however, tells a different story. It suggests that at the most basic biological level, context is just as vital as genetics for establishing the conditions under which potentiality becomes actuality. What's more, it suggests that as environmental conditions change, biological organisms can potentially develop in multiple ways.

Prior to 1997 the cells of the inner blastocyst seemed to have only one potential: to develop into a fetus. This potential seemed naturally given in the sense that "this is what nature does," the cells of a blastocyst develop into the cells of a fetus. However, When Jamie Thomson and his lab team removed the inner cell mass cells from the blastocyst and placed these cells into different media (i.e., a different chemical environment), they discovered something quite astonishing. The cells did not die. Rather, they exhibited a potential that would not have otherwise been discovered: these same cells, with their same genetic program, could become something quite different—stem cell lines. Put more technically, by placing these cells into a new context, researchers reworked the genetic signaling pathways that direct the form and function of the cells. Through a change in context, these cells were given new forms, new functions, and new capacities. Genetics alone, as crucial as they are, could not have predicted let alone programmed this. Natural potentials are dependent for their actualization on their relationships, and not just on intrinsic properties.

Some might argue that stem cell researchers have not discovered new natural potentials, but rather that they stymied or even violated existing potentials. But this argument is not willing to accept that stem cell science teaches us something new about the nature of nature: the potentialities of living systems are not pre-given, capacities are not pre-given, and thus the ends to which these systems can be directed cannot be determined in advance! As such, any appeal to pre-given natural capacities as an argument against stem cell research—such as the pre-given potentiality of the genome—would seem to be unpersuasive. To repeat what was said above, Thomson and subsequent researchers have demonstrated that zygotic cells, under specified conditions, do in fact have capacities other than developing into fetuses, i.e., they can be made to become embryonic stem cells. That these conditions are engineered does not render the altered capacities unnatural. The hES cells do not violate nature. Rather, they exemplify its flexibility, context dependence, and the mutually constitutive relation among innate potentials, contexts, forms, and functions.

Of course this insight into the nature of nature does not settle the question of whether or not embryos should be destroyed, their inner cells harvested, and stem cell lines created. The question of whether or not any of this is good remains unsettled. To quote an apt point from the Vatican's teaching *Donum Vitae*, it is certainly the case that "what is technically possible is not for that very reason morally admissible."[27] It is also the case, however, that the terms of what is admissible cannot be settled by appeal to innate and naturally given potentials, genetic or otherwise.

Does this mean that an examination of nature has nothing to contribute to the ethics of the ethicists? Far from it! Any sufficient evaluation of the ethics of stem cell research must take account of how biology describes the natural world! What biology tells us is that natural systems have multiple potentialities, potentialities that depend on context and form. The question of what forms and contexts are good, therefore, cannot be settled by appeal to nature alone. Ethics is not simply about protecting the unfolding of genetic potentials. Nature can become multiple things; it does not "tell" us what is right and wrong in any straightforward way.

The lessons learned from stem cell research about natural potentialities are, in fact, consonant with the best of Roman Catholic theology. In classical Roman Catholic thought "nature" was never understood as simple biology. Biology alone could never define the terms of "natural law." Natural law is about God's intentions. While these intentions may be reflected in "nature," such reflection could not be reduced to "mere" biology. Biological insights into the nature of nature are not the final word on what things really are. What's more, these certainly cannot be taken as determinative for what things should be. However, what we learn from biology about the nature of nature is deeply consequential: potentialities are not reducible to genetics or other intrinsic properties alone. In the end, even if we never get therapeutics out of stem cell research, we now know a truth that we might not have otherwise known.

So, if it is true that just because something is technically possible, it is not therefore morally admissible, and if it is true that nature is not defined by pre-given potentialities, then how should the question of moral admissibility be settled? We argue that it should be settled through a consideration of future abundance, future wholeness. The ethical question then becomes not: what are the intrinsic potentials of nature, but what constitutes an abundant life? We are certainly not alone in posing this question. Indeed, the question is posed in the introduction to *Evangelium Vitae* itself! We are convinced that the answer to this question does not lie in an examination of the intrinsic nature of things alone, but in a vision of God's promised and our hoped for future.

Conclusion

"The voices of religion are more prominent and influential than they have been for many decades," write the editors of *Nature* magazine. "Researchers, religious and otherwise, need to come to terms with this, while noting that some dogma is not backed by all theologians."[28] The public policy debate over scientific research is at root a theological debate regarding the grounding of human dignity. Because the questions raised by new science and new technology were not anticipated when the founding texts of the great religions were written, the Bible (or the Torah or the Qur'an or the Upanishads) cannot be appealed to for literal answers. Interpretation and reasoning and judgment are called for. Theologians debate among one another in churches and universities, just as vested interest groups debate in laboratories and the public sphere.

The good news is that ethics is now a matter of serious public discussion, and the views of theologians are welcome in many quarters where previously eschewed. The not-so-good news is that we find ourselves in an ethical stalemate and cultural disarray. Our era calls for creative theology and creative ethics right along with creative science.

Notes

1. Benedict M. Ashley, O.P., and Kevin D. O'Rourke, O.P., *Health Care Ethics: A Theological Analysis*, 4th ed. (Washington, DC: Georgetown University Press, 1997), 235.

2. The term "hermaphrodite" combines the names of two Greek gods, the male Hermes with the female Aphrodite.

3. Chimerism may eventually have implications for criminal prosecutions. Courts these days seem to rely increasingly on the unquestioned scientific veracity of DNA testing. If police forensics can match the DNA of the suspect with blood or semen or other bodily evidence left at the crime scene, this seems conclusive for a verdict. Similarly, if a police forensic team finds separate genetic codes in the evidence and in the suspect, the suspect is considered exonerated and frequently acquitted. However, the matter may not be so simple. If the suspect is a chimera, the genetic code of the blood left at the scene may not match the genetic code in semen or other cells. DNA testing could become more complicated than is presently assumed. Both the Vatican and our forensics laboratories should take note.

4. Congregation for the Doctrine of the Faith, *Declaration on Procured Abortion*, in Acta Apostolicae Sedis 66 (1974), note 19.

5. Congregation for the Doctrine of the Faith, *Instruction on Respect for Human Life in Its Origins and on the Dignity of Procreation (Donum Vitae)* (February 22, 1987), Acta Apsotolicae Sedis 1988, I, 1.

6. *Donum Vitae*, I, 1.

7. John Paul II, *Evangelium Vitae* (March 25, 1995), 60, 468–69.

8. Norman Ford, "The Human Embryo as Person in Catholic Teaching," *National Catholic Bioethics Quarterly* 1.2 (Summer 2001): 158.

9. Ford, "The Human Embryo," 160, Ford's italics. See Ford, *When Did I Begin?* and his more recent work, *The Prenatal Person: Ethics from Conception to Birth* (Oxford, UK: Blackwell, 2002). In California, CIRM grantees find their guidelines limit their use of early embryos to 12 days, rather than 14.

10. William Saletan argues that the 14th day is unnecessarily arbitrary, and that ethical reasoning should permit the harvesting of stem cells well beyond this threshold. "The Organ Factory: The Case for Harvesting Older Human Embryos," www.slate.com/toolbar.aspx?action=print&id=2123269.

11. Norman Ford, "Embryo Research, Cloning, and Ethics," *Chisholm Health Ethics Bulletin* (Spring 2002): 6.

12. Thomas Shannon, "Human Embryonic Stem Cell Therapy," *Theological Studies* 62 (December 2001): 811–24.

13. See Thomas A. Shannon and Allan B. Wolter, "Reflections on the Moral Status of the Pre-Embryo," *Theological Studies* 51 (December 1990): 603–26.

14. Lisa Sowle Cahill, "Stem Cells: A Bioethical Balancing Act," *America* (March 26, 2001), 2, online at www.americapress.org/articles/cahill-stem.htm.

15. Benedict Ashley and Albert Moraczewski, "Cloning, Aquinas, and the Embryonic Person," *National Catholic Bioethics Quarterly* 1.2 (Summer 2001): 196.

16. Ashley and Moraczewski, "Cloning," 194.

17. *Donum Vitae*, I, 6.

18. José B. Cibelli, Robert P. Lanza, and Michael D. West, with Carol Ezzell, "The First Human Cloned," *Scientific American* 256 (January 2002): 44–51.

19. E. S. Revazova, et al., *Cloning Stem Cells* 9, doi:10,1089/clo.2007.0033; cited in David Cyranoski, "Activated Eggs Offer Route to Stem Cells," *Nature* 448 (2007): 116.

20. Mark S. Latkovic, "The Science and Ethics of Parthenogenesis," *National Catholic Bioethics Quarterly* 2.2 (Summer 2002): 253–54.

21. Chad A. Cowan, Joxelyn Atienza, Douglas A. Melton, and Kevin Eggan, "Nuclear Reprogramming of Somatic Cells after Fusion with Human Embryonic Stem Cells," *Science* 309 (August 2005): 1369–76.

22. K. Takahashi and S. Yamanaka, "Induction of Pluripotent Stem Cells from Mouse Embryonic and Adult Fibroblast Cultures by Defined Factors," *Cell* 126.4 (August 2006): 652–55. See "Simple Switch Turns Cells Embryonic," *Nature* 447 (June 2007): 618.

23. Dieter Egli, Jacqueline Rosains, Garrett Birkhoff, and Kevin Eggan, "Developmental Reprogramming after Chromosome Transfer into Mitotic Mouse Zygotes," *Nature* 447 (June 2007): 679–85.

24. See Richard M. Doerflinger, "Washington Insider," *National Catholic Bioethics Quarterly* 7.3 (Autumn 2007): 458. See also Nicholas Wade, "Biologists Make Skin Cells Work Like Stem Cells," *New York Times* (June 7, 2007), online at www.nytimes.com/2007/06/07/science/07cell.html?ex=1181880000&en=761ce90dca1b449a&ei=5070&emc=eta1.

25. Insoo Hyun, "Stem Cells from Skin Cells: The Ethical Questions," *The Hastings Center Report* 38:1 (January–February 2008), 20.

26. Hyun, "Stem Cells from Skin Cells," 22.

27. "Introduction," in *Donum Vitae*.

28. "Where Theology Matters," *Nature* 432 (December 2004).

~

CHAPTER TWELVE

The Vatican Argument
in a Cracked Nutshell

In previous chapters we have sought to outline the embryo protection position sponsored by Vatican moral theologians, and we have pointed out some of its problems and inconsistencies. This included some of the problems that knowledge gained from modern medical science creates for Vatican assumptions about biology. In this chapter, we expand our analysis, combining an internal critique with an external critique. With the external critique, we believe we can actually support regenerative medicine including human embryonic stem (hES) cell research from within the embryo protection framework.

Introduction: Every Human Life Is Precious

"Every life is a priceless gift of matchless value," emphasized President George W. Bush during a May 23, 2005, press conference. There is "no such thing as a spare embryo," he reiterated. Embryos are not raw material to be exploited. While President Bush is not Roman Catholic, his stance reflects the embryo protection stance that we examined in the last chapter. Like the Vatican, he believes that the early embryo at the blastocyst stage possesses morally protectable dignity. He opposes destruction of blastocysts for derivation of stem cells. But his arguments, like those of the Vatican, ought not be taken as warranted without closer inspection. Nor ought their resounding "no" to be taken as the only religiously founded position in the public stem cell debate.

Is it possible to begin from an embryo protection framework and nonetheless *support* stem cell research? We believe it is. In this chapter, we will present our own argument within the embryo protection framework to *permit* rather than to *oppose* human embryonic stem cell research. We will open by describing again but in brief the position taken by those who oppose human embryonic stem cell research. We will try faithfully to explicate this argument with its scriptural and theological support, summarizing many of the things we have previously said in more detail.

Then, in Thomistic fashion,[1] we will take a turn toward our own position, offering mild yet what we believe to be significant refutations. We will applaud the diligence of embryo protectionists for defending human dignity; yet we will argue that there are other ways to understand appropriate "protection" for the blastocyst from which we derive stem cells. We believe one can safely protect human dignity while still saying "yes" to stem cells.

Because the existing theological arguments in opposition to stem cell research appeal to the presence in the early embryo of a spiritual or immortal soul, the concept of the soul within Christian theology becomes relevant. We will therefore engage in some soul talk, though we save the bulk of that for a later chapter. We find that the embryo protection position tacitly assumes a "substance dualism," that is, a position in which the person is thought to consist of two intertwined substances, the body a material substance, and the soul a spiritual substance. We, on the contrary, will suggest that the soul is better understood within the larger context of the human person as inclusive not only of body, soul, and spirit but of relationships as well; and what is decisive is that dimension of our human reality that overlaps with the life of God. We have dignity because God calls us into relationship with the divine life.

We will not present here a fully developed alternative theological position, although we will offer the rudiments of what we believe. The burden of proof should be shouldered by embryo protectionists who raise theological arguments against the advance of regenerative medicine. For embryo protectionists to argue on the basis of theology that blocking an opportunity to improve human health and well-being is the right thing to do requires overwhelming evidence. It is tantamount to saying that it is God's will to let millions of people continue to suffer from genetically related diseases and traumas. To justify passing by on the other side while suffering continues will require a very persuasive argument. We are not persuaded. To that unpersuasive argument we now turn.

Is the Soul an Immortal Substance?

In Thomistic fashion, we begin with a query: *Is the soul an immortal substance created by God and imparted to an individual person at conception?* As noted in the previous chapter, if we listen carefully to what is being said by the Vatican along with some American and British Evangelicals who oppose hES cell research, it would seem that at the moment of conception God creates a new immortal soul and imparts it to an individual person; and the presence of this soul in the early embryo renders it at the blastocyst stage morally protectable from destruction in laboratory research. It would seem, then, that the destruction of the blastocyst to derive stem cells would be tantamount to an abortion. An abortion is considered to be the murder of a human person, and therefore the destruction of the blastocyst for purposes of deriving stem cells is also murder.

At first glance, the Bible seems to support the idea that God imparts an immortal soul at conception. In the prophets we read that Israel's divine redeemer "formed you in the womb" (Isaiah 44:24). The Psalmist addresses God: "For it was you who formed my inward parts; you knit me together in my mother's womb" (Psalm 139:13). Such passages have supported arguments against abortion in the past and they can be mustered to support opposition to stem cells now. If God "forms us" or "knits us together" in the very womb, then it seems that God gives us an immortal soul at that point.

Indeed, theologians seem to agree that God imparts souls to individuals. St. Thomas Aquinas asks whether two or more persons can share a soul, and he answers in the negative. "It is impossible that one individual intellectual soul should belong to several individuals."[2] This classical commitment directly applies to the contemporary discussion. In his fine study on the history of the theological understanding of embryo ensoulment, David Albert Jones writes, "The embryo is clearly a living being, an individual and is human."[3] This means the early embryo is an individual person, and this renders it ready to receive the divinely imparted soul.

But when does the individual receive this soul from God? Again at first glance, it makes sense to argue that this individuality is established at the moment of conception, when the genetic code of the woman and the genetic code of the man establish for the first time a new and unique genetic code. What better sign of autonomous identity could nature provide than a novel genome? This, philosophically speaking then, is the moment of individuation, the moment of personhood, the moment of ensoulment. Pope John Paul II says that "modern genetic science offers clear confirmation . . . [that] from the first instant there is established the structure (*structuram seu*) or genetic program of what this living being will be: a man (*hominem*), and indeed this individual man (*hunc hominem individuum*)."[4] Even if we cannot scientifically date the precise moment of ensoulment, we must philosophically date it at conception in order to be safe in our judgment.

Pope John Paul II clarified the logical connection between the presence of the soul and morally protectable dignity. "It is by virtue of the spiritual soul that the whole person possesses such a dignity even in his or her body. . . . If the human body takes its origin from pre-existent living matter, the spiritual soul is immediately created by God."[5] The contrast between preexistent matter and the addition of an ontologically distinct spiritual soul suggests that the soul is of a different substance than the physical body. It is spiritual, not material. This presence of a nonmaterial substance is what justifies morally protectable dignity.

Donum Vitae similarly connects soul with dignity and protection: "respect, defense and promotion of man, his primary and fundamental right to life, his dignity as a person who is endowed with a spiritual soul and with moral responsibility, and who is called to beatific communion with God."[6] The Center for Bioethics and Culture, which garners support from the Evangelical community in opposition to both abortion and stem cell research, adheres to the same logic. "The ethics of abortion, reproductive technologies, genetic manipulation and contraception are all closely tied to when a human embryo is recognized as having significant dignity . . . the presence of a soul, as the seat of the *imago Dei*, renders that dignity as that which separates humanity from the rest of creation."[7]

Because the impartation of a spiritual soul is a divine act associated with the establishment of an individual human person with a unique genetic code at conception, all of what has been said applies regardless of whether the conceptus is in the mother's body (*in vivo*) or in the laboratory (*ex vivo*). It is *ontologically* the case that the zygote possesses a spiritual soul and, thereby, possesses morally protectable dignity. The presence or absence of nurturing surrounding conditions, such as the mother's body, are irrelevant to the embryo's autonomously established ontological status. Similarly irrelevant are any specific characteristics of the developing embryo.

Furthermore, this ontological status is buttressed by the observation that all totipotent cells have a *hypothetical potential* for becoming a human person, even *ex vivo*. Whether totipotent embryonic cells exist in a suitable *in vivo* environment for their potential to become actualized is irrelevant to their moral status.

With this, the logic of the tacitly sacred cell becomes clear. The blastocyst is deemed to be sacred because it either possesses or is about to receive a spiritual soul. Because of this sacred status, it is inviolable. It is morally protectable. It has dignity. It should be treated as an end and not a means. Its life should be spared. The blastocyst becomes sacred and untouchable.

Putting all these considerations together, then, it is morally illicit to destroy the *ex vivo* embryo at the blastocyst stage to harvest stem cells. This is the basic argument from within the "embryo protection" framework that has so influenced public policy. But is it the best or the only possible theological interpretation of what happens in embryo development and its moral implications? We think not.

Person-in-Relationship

Our alternative argument relies upon a relational understanding of the human soul, not on a substantialist metaphysics. We place the soul in dialectical relationship with the spirit. On the one hand, the soul is our centered self, our identity, so to speak. On the other hand, in the spirit we are in relationship with other souls and with God. Rather than connect personhood to a nonphysical substance, we work with a relational or holistic anthropology that connects personhood to body, soul, spirit, and community.

We are not alone in this. Hints of this position can be found in the thought of Pope Benedict XVI as well. We note that the former cardinal Joseph Ratzinger, now Pope Benedict XVI, says, "A being is the more itself the more it is open, the more it is in relationship."[8] This stresses the significance of relationality. It implies a nonsubstantialist understanding of the human soul.

We concur with this stress on relationship. With this in mind, we argue that every person is a person-in-relationship, and this includes a life-nourishing relationship to both God and our parents. Our spiritual and immortal destinies are contingent upon God's action, calling us to communion within the divine life. We adhere to this relational view of the human person rather than viewing the person as possessing a soul with a spiritual substance that adheres to a unique genome. We interpret the New Testament description of the human person as inclusive of body, soul, and spirit; we do not conflate spirit with soul.

We therefore recommend that a theology of the soul begin with our appointed destiny as human persons—that is, our destiny to live in communion with God's Trinitarian life. It is the call forward to our resurrection into the new creation that retroactively determines the dignity we now enjoy. When God names us or calls us from our mothers' wombs, it is the future to which we are being summoned. It is God's valuing of us in this call that we affirm when we impute dignity to ourselves and to the least of God's creatures in the present time.

This theology of the person does not appear to us to have direct application to the question of stem cell research. Certainly, indirectly, it affirms the human enterprise of medical research, according to which we pursue improvement in human health and well-being. To overcome suffering now anticipates God's promise for the New Jerusalem where "mourning and crying and pain will be no more" (Revelation 21:4). But, as we have said earlier, this belongs to the future wholeness framework. Here our concern is not with that framework but with the embryo protection framework and whether it can be used to support stem cell research.

The opposition to stem cell research that emerges from embryo protectionists is laudable to the extent that it seeks to lift up the dignity of all human life, even and especially life that is weak and vulnerable and cannot defend itself. The fundamental moral motivation of Vatican Catholics and Evangelicals to protect the dignity of the weaker among us is healthy. It is godly. It is worth applauding.

Yet, we believe this diligence is misplaced in the stem cell debate. The blastocyst *ex vivo* is not the same as the embryo *in vivo*. Both are valuable and both deserve a level of honor and respect. Certainly we would not argue that the early blastocyst is just "so much tissue" for any and all kinds of experiments. However, it must be clear that its destiny cannot be the same as the destiny of a blastocyst in the womb. The blastocyst in the womb has the potential to become a child and then an adult. Whether it will is, of course, dependent on many things, such as the woman's decision to keep it, the way society protects and supports the infant or fails to do so, and so on. The blastocyst outside the womb has the same potential *only if* it is implanted. Both are contingent on a coexistent life in a mother's body.

In a relational view, in other words, the question of environment and implantation becomes crucial. The normal destiny of the embryo *in vivo* is to become a child. The normal destiny of the blastocyst *ex vivo*, however, is different: If it is not implanted, its destiny is not childhood but death. We argue that this different destiny is morally relevant. We are not the only ones to think this, either—other ethicists have argued that there is a relevant difference between the blastocyst outside the womb who will die in the normal course of events and the embryo in the womb who will live in the normal course of events.

Now some might argue that it is our moral responsibility to provide for every blastocyst the possibility of becoming an adult. All fertilized embryos should therefore be implanted. This would be a logical extension of the moral reasoning that takes "ensoulment" to be what provides protectability, and

that argues for ensoulment at the moment of conception. But, to attach ensoulment to genomic novelty at conception is unwarranted, in our view. The
soul is not a metaphysical substance injected by an angelic hypodermic needle
into a zygote. Rather, the soul is the product of our dynamic relationship with
God that develops within each of us a profound inner self and sense of self.

Many passages in the Bible support our relationalist view. They dramatically demonstrate the care and attention that God has for each one of us,
calling us from nonbeing into being and finally into fellowship within the divine life. "The Lord called me before I was born," says Isaiah, "while I was in
my mother's womb he named me" (Isaiah 49:1). What gives the prophet and
the rest of us dignity is God's call, God's naming us as God's own. Nothing is
said here about our possession of a spiritual or immortal substance known as
the soul. What is decisive here is our *relationship* to God, a relationship God
establishes by calling us toward the divine.

Furthermore, the place where this divine call takes place is the mother's
womb. Nothing is said here about conception at all, let alone *ex vivo* conception. Nothing anywhere in the Bible would support the highly philosophical
idea of a spiritual entity known as a soul specially created and then imparted
at conception. It is understandable how such passages could be enlisted to
support opposition to abortion, when abortion is understood as removal of
a baby-to-be from a mother's womb. They do not say enough, however, to
establish any ontology of an individual soul.

When St. Thomas Aquinas emphasized that a soul is for an individual,
he had in mind the rational faculty of the soul. He understood the soul as
intellect. Yet Aquinas was not a substance dualist in the tradition of Plato
or Pope John Paul II; rather, more like Aristotle, he viewed the soul as the
form of the body, as intellect embedded in the body. Further, his focus on
intellect would seem to preclude the earliest of embryos. Indeed, Aquinas did
not believe that intellectual ensoulment happened until considerably later
in pregnancy.

The concept of soul with which contemporary Vatican ethicists work
emphasizes less the intellectual faculty and more the spiritual nature, its immortality. Neither Cardinal Ratzinger as reflected in *Donum Vitae* nor Pope
John Paul II as reflected in *Evangelium Vitae* would attribute the capacity
for reasoning to the freshly fertilized zygote. It does not have a brain yet,
let alone a mind. Perhaps we have a problem of equivocation at work here.
Are we operating with multiple definitions of what a soul is? One can only
ask: Where does this contemporary emphasis on biological individuality and
spiritual individuality come from?

One might argue that the early embryo may not yet possess intellectual capacity, but it would have intellectual potential under the right circumstances. However, such potential need not be individual potential; it could be common human potential. This is what Roman Catholic ethicist Thomas Shannon argues. As we report in the previous chapter, potency is not the same thing as actuality, he says. A potential person is not yet a person, nor is a potential intellect an actual intellect. Shannon can concede that there is a common human potential present in the early embryo, but not a potentiality belonging to an individual person.

Our only point here in this reply is that we will concede, for the sake of argument, that we should associate the human soul with a human individual person. We will not contest this assumption. The question we next turn to is this: When in early embryo development can we speak of an individual person?

Soul and Dignity

What has not been made clear by embryo protectionists is the logical connection between the alleged presence of a spiritual or immortal soul, on the one hand, and morally protectable dignity, on the other hand. This seems decisive in Vatican arguments; yet, nowhere is this connection spelled out. Indeed, embryo protectionists fail to ask just what is encompassed under an assumption of personhood for the early embryo. At least one of the authors of this book takes the view that the embryo from its earliest development can be called a "person," but that the implications of this need not mean that it must be protected from stem cell research. The typical protectionist argument seems to presuppose that if a "spiritual soul" or moral personhood is present, then life must be preserved; and if it is absent, then life may be taken.

We have several problems with these correlating presuppositions. First, this view presumes, as noted above, a substance dualism: the soul is an entity made up of a spiritual substance that stands in distinction to the material substance that composes our bodies. Despite the Vatican's sustained emphasis on the "whole person as body and soul, " the assumption of substance dualism is clear in what Pope John Paul II writes.

As noted, we embrace a much more relational anthropology, wherein we understand the soul and the spirit in terms of our relationship with God. It is the relationship we enjoy with God that affords us our dignity and our eternal destiny. Rather than ask our unique genomes to attach themselves to a unique soul, we believe our entire existence—our entire life stories—are enveloped within the larger history of God with our creation.

For support we enlist the theology of former Joseph Cardinal Ratzinger, now Pope Benedict XVI. "The challenge to traditional theology today lies in the negation of an autonomous, 'substantial' soul with a built-in immortality, in favor of that positive view which regards God's decision and activity as the real foundation of a continuing human existence."[9] It is our relationship with God that determines the benefits we normally associate with the soul. Our Trinitarian God "is relationship, since he is love. It is for this reason that he is life . . . relation makes immortal; openness, not closure, is the end in which we find our beginning."[10]

Second, if individuality is requisite for ensoulment and for morally protectable dignity, then one would expect to see this dated at implantation. One would expect to incorporate into the moral calculus the relation of the embryo to its mother. Further, one would expect to see the significance of the neuronal ridge honored, as it is by many Roman Catholics. In short, if individuality matters to ensoulment, then ensoulment cannot be said to occur at conception. Given their own assumptions, embryo protectionists should make implantation decisive. One can work within an embryo protection framework and still find it acceptable to use the blastocyst for stem cell research.

Third, the strictly hypothetical potential for each totipotent embryonic cell to develop into a human being is insufficient for grounding a concrete moral decision or to set public policy. Some differences are morally relevant. For instance, we require "informed consent" before treating adults, because they are autonomous individuals and therefore their decision-making must be respected. This is a time-honored tradition in medicine with many good arguments to support it. But just as surely, we do not require informed consent from children, for they are not yet autonomous and not able to make reasoned decisions about what is in their best interests. Similarly, the early embryo is not autonomous. More importantly, it has no capacity for pain. Animals are not autonomous, yet they have the capacity for pain and therefore may not be treated in a cruel fashion. Without the capacity for pain, the question of cruelty does not apply.

There can be no constraints against stem cell research, then, based on autonomy or on concerns about cruelty. The only remaining moral question is whether the life alone of the early embryo must be preserved. Is it morally wrong to end the life of the blastocyst? Here, one can take the typical embryo protectionist view that it is morally wrong to do so. But one can also argue differently. The blastocyst *ex vivo* is destined for death. By continuing its existence as a line of stem cells, one in fact prevents that death and gives it a form of ongoing life. To "respect" it, in other words, may mean precisely to keep it "alive" in the form of stem cells.

Conclusion

For many reasons we argue that it is possible to remain within the embryo protection framework and still support regenerative medicine, including stem cell research.

First, we note that the contemporary maxim—to be a person is to be a person-in-relationship—is no more dramatically illustrated than in what happens at implantation and the appearance of the primitive streak. The very individuality of the embryo becoming a fetus is dependent on its relationship to the mother's body. The woman's hormones signal which genes within the fetus will express themselves. It takes two for the baby-to-be to become one.

Without implantation and sharing in the woman's life, no potential for human development exists. Any philosophy of the early embryo that tries to establish an autonomous ontological status independent of its relationship to the woman and to the circumstances necessary for growth and development is without biblical warrant and lacks credulity in the face of contemporary embryology.

To say that the *ex vivo* blastocyst possesses an individual spiritual soul that God had previously imparted seems arbitrary, a supposition that goes well beyond what is justifiable by scripture, tradition, reason, or experience.

Second, the tacit logic of the embryo protectionist is to perpetrate a confusion regarding what is sacred. We believe God and God alone is sacred. We further believe that human persons have dignity—that is, each person should be treated as an end and not merely as a means. At an ethical level, we treat each person as if he or she were sacred, while recognizing that each of us is but a creature of the holy God. The blastocyst in the laboratory does not possess either sacredness or dignity. The early embryo is special, to be sure; because it represents the human life force. Yet, in itself it is not divine; nor is it the soul of an individual human person. We find the arguments of the embryo protectionists to be confusing on these points.

Third, we note an implication of the Vatican position that tends toward a *reductio ad absurdum*. If one is totally dedicated to preserving the life of the embryo, then one must *either* argue that all embryos should be implanted into the womb and be brought to birth *or* one must recognize that creating stem cells provides a way for an embryo otherwise slated for death to still give of its life for the life of others. It would be absurd to demand that every fertilized egg become implanted and brought to birth, because nature herself jettisons so many and because far more fertilized ova sit in IVF freezers than there are women to bear them. It is simply a fact that implantation of all early embryos is unrealistic. So, one might ask: If an *ex vivo* embryo has no potential for becoming a born baby, might there be ethical grounds for sharing its life-giving potential for the health and well-being of future stem cell recipients? With these observations, we believe it is possible to be an *embryo protectionist* and still support regenerative medical research. Note that we are saying more here than we said earlier. We contend that a full-fledged embryo protectionist can provide moral warrant for supporting this area of medical science. Indeed, a true embryo protectionist would prefer the development of stem cells to the simple discarding of the *ex vivo* embryo.

We recognize that we have not yet provided a full-fledged theology of the human person, inclusive of soul as well as spirit. We have not answered all the important questions regarding human nature. We have not provided what so many ask for, namely, a marker on the calendar of human development where inviolable human dignity begins and life cannot be destroyed. Although we feel just as zealous regarding the protection of human dignity as the embryo protectionists, we believe the typical arguments applying this zeal to the *ex vivo conceptus* are inadequate. We believe we have demonstrated this.

We have shown here that those who are committed to shutting down stem cell research on the grounds of opposition to the destruction of the early blastocyst are using a weak argument. The case for absolute inviolability of an individual person has not been made. Anxiety over the fate of newly created souls is needless. Worry that our society is violating God's intention for precious human life is unfounded.

In these last several chapters, we have examined the best arguments put forth to date, and have found them wanting. They do not sufficiently persuade. The burden of proof lies on the shoulders of those who would shut down medical research. Those who support medical advance need not ordinarily justify themselves. The potential of regenerative medicine for dramatically improving human health and well-being is virtually a self-validating moral concern. *Prima facie* it embodies an inherent good. Therefore, any argument that would result in retarding or eliminating such medical advance must be overwhelming. It must be convincing beyond a shadow of a doubt. The argument put forth by the Vatican and other embryo protectionists falls short of making a convincing case. Until a better argument against benefiting the human race through medical advance is made, the theologians authoring this book support stem cell research.

As theologians, we would like to follow this chapter with a more constructive discussion of the human soul along with person-in-relationship. We will address this, eventually. But, before we address the matter of the soul directly, we wish to continue our analysis of the human protection framework while listening to one of the dominant voices in the public debate, Leon Kass. We will also look at other topics within the larger public debate, turning subsequently to the question of the human soul.

Notes

1. Ted Peters originally crafted this entire chapter in true Thomistic fashion, after St. Thomas Aquinas's definitive work of the twelfth century, *Summa Theologica*. As scholastic style is not easy to read, however, we have modified that approach here. In good Thomistic form, nonetheless, we begin by trying to present as strong an argument as possible for the position that one wishes eventually to refute.

2. Thomas Aquinas, *Summa Theologica*, I:I:Q.76.A.2.

3. David Albert Jones, *The Soul of the Embryo* (London and New York: Continuum, 2004) 224.

4. John Paul II, *Evangelium Vitae* (March 25, 1995), 60, 468.

5. Pope John Paul II, "Evolution and the Living God," in *Science and Theology: The New Consonance*, ed. Ted Peters (Boulder, CO: Westview, 1998), 151.

6. Congregation for the Doctrine of the Faith, *Instruction on Respect for Human Life in Its Origins and on the Dignity of Procreation (Donum Vitae)* (February 22, 1987), Acta Apsotolicae Sedis 1988, I., 1.

7. Steven Suits, "Ensoulment and the Sacredness of Human Life," The Center for Bioethics and Culture Network, www.thecbc.org/redesigned/research_display .php?id=39.

8. Johann Auer and Joseph Ratzinger, *Eschatology: Death and Eternal Life*, vol. 9 of *Dogmatic Theology* (Washington, DC: Catholic University of America Press, 1988), 155.

9. Auer and Ratzinger, *Eschatology*, 150.

10. Auer and Ratzinger, *Eschatology*, 158.

CHAPTER THIRTEEN

~

Leon Kass,
Protector of Human Nature

During the summer of 2001, George W. Bush alerted the world that he was cogitating about cloning and stem cells. He consulted with two prominent individuals, Pope John Paul II and Leon Kass. He went to Rome to see the pope; he invited Kass to the White House. The pope represents the embryo protection framework; Kass represents the human protection framework.

The president said to the highly respected University of Chicago professor of social thought, Leon Kass, "Get to work." On January 17, 2002, that work began with the first meeting of the President's Council on Bioethics (PCBE).[1] The agenda Kass set for that first meeting introduced a series of concerns that would serve to structure and orient the remainder of the Council's deliberations and publications.

Kass opened with a foundational question: "How ought we to do bioethics and do it well?" Bioethics done well, Kass proposed, "must always attend to the deep character of human individual and social *bios* and how they interact with findings of biology and the technical powers they make possible." Accordingly, said Kass, the Council's principal work would consist of an ongoing project "to develop attitudes, ideas, and approaches for a richer and deeper public bioethics, one that does justice to the full human meaning of biomedical advance."[2] This ongoing project would prove the measure of the PCBE's identity and concern.

Kass brought to Washington his own ongoing project and made it the nation's project. His own position stems from his view of the foundational stakes of bioethics: Kass thinks that biotechnology in the contemporary world poses a threat to human nature and human purpose. Biotechnology threatens to dehumanize. Our ethical mandate is to protect our human nature from dehumanization.

Meeting only six months following the 9/11 attacks on the Pentagon and World Trade Center, Kass placed the problems of bioethics in the arena of national security. Thus for Kass, bioethics is a matter of security: The "humanness" of being human must be secured against the potentially dehumanizing force of biotechnology.

Part of Kass's previous project had been to argue that the field of bioethics, generally speaking, is not up to this task. And so, Kass's principal and persistent concern for the Council was the development of bioethics commensurate with this perceived need for securing humanness against the dangerous and seductive advances of science. He thinks of this as a "richer bioethics"—one that seeks to do justice to the full human meaning of biotechnological advance." What Kass had previously thought in Chicago would now radiate from Washington to all of America. To the extent that Kass was successful in establishing such a richer public bioethics, the value of his contributions must be underscored.

Human Dignity and Human Nature

The field of bioethics was born little more than 35 years ago, and Leon Kass was one of the midwives.[3] In a 1972 essay, Kass posed questions that subsequent bioethicists have now inherited: Are we playing God? What means are acceptable to reach desired ends? What is the significance of separating sex from procreation? Is there a right to a unique genome? What risks should we take on behalf of children? Are women and their bodies being exploited in exotic reproductive technologies? Are children threatened with commodification? What is the significance of shifts of power into the hands of a few people? Perhaps most important on the list: Are we being dehumanized?[4]

From Kass's perspective, the bioethical challenge is to keep us genuinely human in the face of threats to dehumanize us coming from biotechnology. Cloning and stem cell research constitute such dehumanizing threats. When bioethicists side with biotechnologists, then the field of bioethics becomes impoverished, perhaps even dangerous. Kass would like to enrich bioethics by incorporating protections that will preserve human nature and human dignity.

In his recent book, *Life, Liberty and the Defense of Dignity*, Kass identifies his focal striving, namely, "to protect and preserve human dignity and the ideas and practices that keep us human."[5] His thesis is that "the new biotech-nologies threaten not so much liberty and equality as something we might summarily call 'human dignity.'"[6] How is our dignity threatened? By adopting and embracing new medical technologies, we might lose appreciation for our inherited human nature. This threat from technology is not external. It is in-ternal. We ourselves might choose it. "Most repulsive," asserts Kass, is that we might choose willfully self-debasement, dehumanization, and degradation.[7]

What does Kass associate with dehumanization? Three things. First, de-humanization occurs when treating children as products of our will rather than as mysterious strangers to be cherished. Children lose dignity as they become commodified. "Procreation dehumanized into manufacture is further degraded by commodification, a virtually inescapable result of allowing baby-making to proceed under the banner of commerce."[8] Assisted reproduction leads to commodification, and this puts us on a slippery slope toward dysto-pia. He affirms natural procreation over technically assisted reproduction.

Second, Kass says we dehumanize ourselves when we try to avoid our "given nature." Our given nature includes finite limits. Disease and death belong on the list of finite limits that adhere to human nature. In addition to limits, we are also beset with a contradictory impulse or aspiration to tran-scend limits. "Human aspiration depends absolutely on our being creatures of need and finitude, and hence of longings and attachments."[9] Kass invites us to stop and ponder and appreciate who we are as finite and embodied human persons, and this appreciation testifies to our dignity. "At stake is the *idea* of the *humanness* of our human life and the meaning of our embodiment, our sexual being, and our relation to ancestors and descendants."[10]

The third force for dehumanization is our desire for rational mastery. It is our aspiration to transcend our limits that gives rise to "the problem of technology." Following Martin Heidegger and Jacques Ellul, Kass believes "technology and technique are not today limited to external and physical na-ture; technology now works directly on the technologist, on man himself."[11] That is, as a culture we have internalized technology so that we see it as self-defining, and this can lead only to the tragedy of self-debasement. With this as his observation, Kass rejects cloning and other assisted reproductive technologies because they cause us to forget our vulnerabilities as embodied creatures with a finite place among living beings.

Sex, Nature, and Ethics

Kass asks nature to provide us with our moral norm. He begins with *description* and then moves to *prescription*. He begins with a description of the existential power of sexuality within marriage, the procreation of children who will live on to future generations, and the salutary way this meets our deep inner need for self-transcendence. Kass prescribes that we accept what he describes.

Kass begins with a description of the human condition. What Kass sees when he looks at human beings is that we seem to be created with a *telos* or goal to transcend our earthly life. We have a "longing for immortality." But, to be realistic, we cannot be immortal. Yet, not all is lost. "The promise of immortality and eternity answers rather to a deep truth about the human soul; the human soul yearns for, longs for, aspires to some condition, some state, some goal toward which our earthly activities are directed but which cannot be attained in earthly life. . . . One possibility is completion in another person."[12] The need we give voice to when asking for life beyond death is a need that can be met existentially in the love of family life and the extension of our purpose in another person, namely, our child. "Biology also teaches about transcendence, though it eschews talk about the soul. . . . I refer not to stem cells, but to procreation."[13] We transcend ourselves through our children. This is the message nature is teaching us. Once we have learned what nature teaches, we should live accordingly.

As we have said, Kass believes that *description* is the key to open the door to *prescription*. How we evaluate something depends on how we describe it. This is nowhere more important than in the relationship between sexuality and procreation as they relate to ethical questions surrounding cloning and stem cell research.

Should we think of reproductive cloning as merely an exercise of individual rights?[14] Is it a way to enhance human health or improve the species? Kass rejects such prospects and suggests that we must begin by describing the wonders of childbirth and the deep meaning of parent-child relations. Sexual reproduction, given us by nature, foretells deep truths about our identity. These truths are not mere cultural constructs. It follows, then, that all cloning, even on a small scale, would threaten identity and individuality. It would transform procreation into manufacture, and violate the "inner meaning" of parent-child relations. The proper framework for ethical discussion begins with a recognition of the "deep truths" that are embedded in our "given nature." These are truths about parents, sex, and children.

Kass becomes dizzily lyrical when waxing poetic about the power and beauty of sex and family life. Sexual procreation is fitting for "higher" forms of life. Sexual reproduction itself is "soul-elevating" and is deeply connected to human erotic longing. The impulse to reproduce arises out of anxiety about death and reminds us that death is a necessary part of life. Reproduction serves our desire for self-preservation and for transcendence. This means that when we sever procreation from sexuality it is "inherently dehumanizing."

This inextricable connection between embodied spirit and bringing children into the world is the root concern for Kass. Any severing of the deep connection between sexual bonding and the generation of new life will of necessity violate the meaning of being human. To bring a child into the world through cloning would ignore our embodied selves and adopt a "technocratic" attitude toward children. Children would no longer be children, but commodities, property, possessions. So strongly does Kass object to the possibility of ever cloning a human child that he proposes that we need a ban on *all* human cloning, including cloning for research purposes limited to the first 14 days of embryonic development. "Prudence," he suggests, "argues powerfully against permitting *any* human cloning."[15] So, we can see the ethical logic here: As Kass sweeps reproductive cloning into the moral dust bin, he sweeps therapeutic cloning right along with it. And, right along with these two types of cloning, embryonic stem cell research disappears into the same disposal bin.

Bioethics, Terrorism, and Bioterrorism

Kass opened the inaugural meeting of the PCBE with a statement of his view of the significance of the Council's work. Though at the time Kass apologized that his opening statement would be only "semicoherent," the statement has proven a herald. In it Kass articulated the fundamental stakes of bioethics as he understands them.

Kass began by reflecting on the significance of bioethics in a world "changed drastically" by the events of September 11, 2001. At first blush this point of departure was not surprising. The nation's, indeed, much of the world's attention was focused on the dramatic attacks perpetrated against the United States. In the wake of September 11 the potential dangers of an advancing biotechnology appeared to many people to be less immediately worrisome than the insecurities propagated by terrorism. For Kass, however, the connection between September 11 and bioethics was not at all incidental; the two shared a deeper logic. For Americans, September 11 revealed the dark vulnerabilities of a contemporary world we thought enlightened and secure. For Kass, biotechnology carries with it a similar set of vulnerabilities. The stakes of a post-9/11 world and the stakes of biotechnology coincide. In both something fundamental is a risk; something fundamental must be secured: our humanity. A key term for Kass in this coincidence of stakes is "way of life." In both terrorism and biotechnology more than mere life is threatened. Life-worth-living is under attack.

Kass turned the Council's attention to the roots of the word *bioethics*. He suggested that for the Greeks the term *bios* did not mean merely "life as such," a quality characteristic of all living beings, but that it referred exclusively to life as humanly lived, "something describable in a 'bio-graphy.'" *Bios*, said Kass, is life lived "not merely physiologically, but also mentally, socially, culturally, politically, and spiritually." Bios refers to a "way of life." Bioethics, Kass concluded, should thus be an ethic that considers life as humanly lived; and biotechnology should be understood to encompass those technologies that affect such a life.

Understood in this way, *bios* conceptually knits together the stakes of terrorism and biotechnology. In both, *bios* is under threat. Both concern a matter of pressing importance: the defense of life—not only "life as such," but life "humanly lived."

Though the events of September 11 turned the public's attention away from biotechnology in the short run, Kass insisted that in the long run they "paradoxically" strengthened the work of bioethics. September 11 brought a new "moral seriousness." This moral seriousness had swept away a "fog of unthinking and easygoing relativism." Now, Americans can see "evil as evil." By putting the public in a moral frame of mind, September 11 prepared us to face the threats of an advancing biotechnology.

Of course, Kass acknowledged, the moral challenges taken up by the Council are not identical to those related to terrorism. In the case of terrorism, evil is more or less easy to identify. In bioethics, by contrast, "the evils we face, if indeed they are evil," are intermixed with the very goods we seek: "cures for disease, relief for suffering and preservation of life." For Kass biotechnology is tragic—while animated by principles important to our self-definition such as "devotion to life," "freedom to inquire," and "commitment to compassionate humanitarianism," biotechnology risks violating the very thing we seek to enrich—our humanity. Understanding the work of the PCBE consists entirely of understanding the terms of this tragedy.

Concluding his opening remarks, Kass insisted that terrorism and biotechnology present us with a single work: "safeguarding the human future." If terrorism represents a threat to our humanity from without—an "antimodern fanaticism and barbaric disregard for human life"—biotechnology risks being a threat from within—a "utopian project to remake humankind in our own image." How ought we do bioethics and do it well? Both the "inhumanity of Osama bin Laden on the one hand and the posthuman Mustafa Mond, Aldous Huxley's spokesman for the *Brave New World*, on the other," challenge us to question of "the good life, of humanization and dehumanization."

Now, we offer the following observation. Whether or not our humanity is at stake in modern biotechnology depends entirely on how such things as "humanity," "biotechnology," and "modern" are understood. The meaning of these terms is not self-evident. What constitutes a "human way or life" or a *bios* could, according to some definitions, actually be enhanced by biotechnological advance. If we think of the human being as *homo faber*—that is, if we think of technical creativity as built in to human nature—then such advances would be an expression of, not a debasement of, our true humanity. We simply point this out, because Kass's assumptions regarding what counts as truly human are not self-evident.

Human Nature, the Present, and the Future

As Kass proposed it, a richer bioethics consists of three elements: (1) it determines which features of human nature are most worth defending; (2) it judges whether or not biotechnology threatens those features; and, if it does, (3) it defends those features through public opinion and public policy. Through the course of the Council's work, these three elements would emerge as three interrelated inquiries. Outlining the structure of these three inquiries will orient us to the content of the Council's deliberations.

The interrelationship between these three inquiries involves a subtle interplay between the natural and the unnatural. When discussing human nature the Council engages what we flag as a *naturalistic* ethic. The reasoning goes something like this: "Nature," unaltered by human hands, is inherently good. As such, human life as it is "naturally given" also is always already good; thus, it is good before we do anything to make it good. Alterations to what is "naturally given" thus entail the risks of "violating" what is good. In many of the Council's arguments "natural" is rhetorically synonymous with good. Conversely, "unnatural" is rhetorically synonymous with evil.

Key to the Council's ethic is the universal character of what is naturally good. That which is naturally good remains good regardless of context or circumstance. Standards for judging what is good and what is not good about biotechnology are not determined by the consequences of that technology. Rather, they are determined by the degree to which biotechnology coheres with what is naturally given.

The Council's ethic is applied primarily to society or culture, not to individuals on their own. Recall Bush's fear that stem cell research encourages a *culture* that devalues life. The aspects of human life worth defending are threatened not by individuals but by society.

Curiously, this attention to the aggregate effects of individual actions functions according to a consequentialist logic. Nature can be directly violated, but violation comes about as the consequence of the aggregate effect of individual factors. The interplay of the naturalist and the consequentialist ethics animate the Council's fear that contemporary society might be moving toward a "brave new world."

Human Nature: "Defining and Worthy Features"

Consistent with its mission the Council takes up the question: "What is significant about being human?" If the "goodness" or "badness" of biotechnology depends on its relationship to the "defining and worthy features of human life," then the Council must begin its work by determining what those features are. The first task of the bioethicist is anthropological—she or he must investigate the nature of human nature.

Taking a broad survey, the Council seems to agree that human nature consists of two defining characteristics. One Council member, Gilbert Meilaender, has described these two characteristics as "finitude" and "freedom." Marked by finitude, humans are defined by an inherent set of limitations. Marked by freedom, humans are defined by the desire to transcend those limitations—an "impulse toward perfection." The balanced tension between these two characteristics is the measure of the "humanness" of humanity. Humans are "creatures in-between," as the Council describes them, in-between finitude and freedom. Meilaender, and now the Council, gives voice to a rich theological tradition by describing the human condition in this way.

The Council suggests that the human life lived well is one in which human limitations are balanced against the drive to overcome these limitations. "Full human flourishing" involves recognizing that the human is a creature whose "limitations are the source of its—our—loftiest aspirations, whose weaknesses are the source of its—our—keenest attachments."[16] This point is vital to understanding the way in which the PCBE makes judgments about biotechnology. According to the Council our limitations are good; without them we would not aspire to lofty things. Likewise our weaknesses are good; they are the condition for possibility of our forming our keenest attachments.

Presumably a way of life that cultivates and maintains a balance of "freedom" and "finitude" can be described as humanizing. By contrast, the suppression of either finitude or freedom—throwing the relationships between them out of balance—would constitute a violation of our humanity and be described as dehumanizing. The task of a "richer bioethics" is to vigilantly safeguard this balance.

Why Should Natural Limits Be Respected?

The Council's writings admit to two answers to the question: "Why should natural limits be respected?" The first is prudential. Nature, "fallible and imperfect," is also highly complex and delicately balanced, the result of "eons of gradual and exacting evolution."[17] As such, any technical intervention risks triggering dangerous unintended consequences.

The prudential answer, however, is less significant for the Council. The second reason that natural limits ought to be respected is that biotechnical interventions risk a temptation to what the Council calls "hyper-agency." Hyper-agency is described as the "Promethean aspiration to remake nature, including human nature, to serve our purposes and to satisfy our desires."[18] This aspiration constitutes a "false understanding of" and an "improper disposition toward" the naturally given world, a failure to acknowledge the goodness and giftedness of the world. True understanding and proper disposition consists of the recognition that nature is not fully ours to master, but a gift to be accepted and served with "reverent awe."

The Council recognizes that its appeal to the goodness of "natural gifts" may be unpersuasive. Nature's "gifts" are frequently undesirable—disease, predation, suffering. And where nature's gifts are more desirable—health, physical coordination, metabolic stability—the "distribution" of those gifts appears unequal. Some "naturally given" human limits, it would seem, ought to be opposed by human innovation. Certainly the Council does not think science should lay down its arms in the fight against disease and suffering. So we must ask the question: "Which natural limits are good and which ones are not?"

The Council suggests that in order to answer such questions we need to look not to the limits of nature generally, but to those limits characteristic of *given human* nature. Only if there is a *given* human nature, the PCBE argues, a "given humanness, that is also good and worth respecting, either as we find it or as it could be perfected *without ceasing to be itself*, will the 'given' serve as a positive guide for choosing what to alter and what to leave alone."[19]

So is there a given human nature? Yes, suggests the Council. It is "limitation." Which limitations? While failing to give a specific answer to this question, the Council suggests that all human goods are in some way inseparable from limits; that is to say, limits serve as the condition for the possibility of our experiencing the good in life. Our greatest achievements in life consist in our having worked to overcome limitations in ability. Our aging and disease-prone bodies limit our life span, thereby encouraging us to form deep attachments and to pursue "the best things in human life." We know happiness and joy, suggests the Council, because we know pain and loss.

The upshot for the PCBE is this: In acceptance of our finitude we recognize that the good life is inseparably tied to limitations. Moreover, we recognize that these limitations are a given part of who we are as humans, and that if we were to do away with these limitations we would cease to be ourselves. As such we should be wary of pursuing biotechnological alterations that might sever the link between our lives and the limits that make those lives good. Thus the reasoning: (1) certain limitations are integral to what it means to be human; (2) those limitations cannot, therefore, be transgressed without humans ceasing to be human; and (3) bioethics, thus, should go about the business of ensuring that those limitations remain unaffected.

Freedom as the Impulse toward Perfection

But, of course, we are not content with our limits. If human nature is characterized by limitations, it is equally characterized by "deep dissatisfaction" with those limits and a desire to overcome them. It is in this sense that the PCBE identifies human limitation as the condition for the possibility of our "loftiest aspirations." Without limitation we would not strive to overcome limitation.

The Council identifies this aspiration to overcome our limitations with the human experience of freedom. We need not accept life as it is given to us. We imagine a life that is different and go about the business of pursuing that life. The Council writes, humans have "extraordinary powers, unique among the earth's creatures, to shape our environment and even ourselves according to our wills."[20] This freedom allows us to alter the human estate, relieving suffering and guarding against the violence of nonhuman nature.

However, this impulse toward transcending limitation tends toward the desire for perfection. Our capacity for freedom becomes a tragic flaw built into our very constitution. Whereas limitations are "naturally given" and therefore good, the drive to overcome these limitations is described by the Council as a "native human desire," a native desire in tension with the "wisdom" of nature. Like Achilles, our desire to overcome limitation can push us to self-destruction. As humans we are finite, mortal. And yet in our dream of perfection we become incapable of stomaching our own imperfections. Our desire to change our limitations becomes a quest to escape all limits. This drive to relieve ourselves of our own finitude is nothing less than our desire to relieve ourselves of our own humanity. In the Council's words, not only do we strive to "kill the creature made in God's image," but we seek to "remake ourselves after images of our own devising."[21] The condition for the possibility of dehumanization is tragically built into our very nature.

The Council is careful to point out that it is not science or technology per se that threaten humanity with tragedy. Rather, it is the way in which they help us realize certain drives and desires. Noting that it is more than "processes and products," the Council writes that biotechnology is "a *conceptual and ethical outlook*, informed by progressive aspirations . . . a desire and disposition rationally to understand, order, predict, and (ultimately) control the events and workings of nature."[22] As such, biotechnology "is a form of human empowerment." In order to judge the desirability of biotechnology we must assess whether the power it provides is informed by respect for our given limitations, or by a desire to overcome them.

Why does the Council worry about all this? Because they think that the "defining and worthy features of human life" cannot be attained strictly through biomedical means. Insofar as our scientized and liberal society reinforces the hope that human fulfillment can be attained through biomedical interventions, individuals within that society will tend to choose those interventions. More importantly, the Council thinks that the pursuit of human betterment through strictly biomedical means will result in the destruction of the very life it seeks to ameliorate. Remember that for the Council the good in life is found in the interplay between life's limitations and our drive to transcend those limitations. The more a medically technologized view of the world teaches us to see all of life's limitations as diseases to be cured, the more likely society is to pursue the transcendence of those limitations by any means. In attempting to overcome these limitations, the Council predicts, human beings will, at best, live an unfulfilled existence, and, at worst will destroy the very conditions for the possibility of their own happiness by destroying all aspiration to "genuine flourishing."

In sum, the "defining and worthy features" of being human adhere in the tense balance between natural limits and the desire to overcome those limits. Kass's proposal for a *richer bioethics* consists in the defense of human limitations against the tragic impulse to perfection that would eliminate those limitations, and, in so doing, eliminate humanity itself.

From *Description* to *Prescription*: A Critique

The naturalistic method for pursuing moral direction follows a path from description of the human condition to prescription in bioethics. This direction of argument is commonly dubbed the *naturalistic fallacy*. What is at stake when we speak of such a fallacy? We ask: Can nature be our guide when it comes to moral norms? Does the attempt to move from nature to ethics constitute a form of the naturalistic fallacy?

Philosopher G. E. Moore, writing on the influence of Social Darwinism, was not the first one to raise the issue or use the term *naturalistic fallacy*, though it is usually associated with him. There are *two* crucial issues. One is whether ethical terms such as "good" or "right" can be *equated* with natural terms such as "conducive to growth" or "admired by everyone." If we say, "that is good," is that statement equivalent to saying "that conduces to growth"? The other issue is whether normative ethical judgments about what is "good" or "right" can be *derived* from "facts"—i.e., whether they can be derived using inductive or deductive methods.[23] For shorthand we ask: Can *describing what is already the case* in nature provide the moral norms for *prescribing what ought to be*? If one believes that the future should be different from the past and that one's moral norms prescribe what transformations ought to take place, then the kind of argument a naturalist raises will seem ethically inadequate.

Here we will look briefly at three relevant examples and offer a critique of each: *nature's blessing on the nuclear family; the wisdom of repugnance*; and *the forbidding of stem cell research*.

The first is Kass's appeal to nature to support the traditional nuclear family with heterosexual congress as a necessary ingredient in human dignity. This means old-fashioned sex is the only moral way to make a baby. Why? Because sexual procreation is the "natural" way of mammalian reproduction. Technologically assisted baby-making such as cloning is "unnatural," then, by virtue of its asexuality. How did we draw this conclusion from that premise?

What Kass considers to be the natural family is the nuclear family that shares an inclusive tradition with its ancestors and descendents. He dismisses without argument feminist and postmodern contentions that the form of the family is socially constructed. "Thanks to feminism and the gay rights movement, we are increasingly encouraged to treat the natural heterosexual difference and its preeminence as a matter of 'cultural construction.'"[24] What some see as socially constructed, Kass sees as grounded in nature. In addition, Kass is willing thereby to universalize the implications of his ethical argument, so that it applies to human beings everywhere. Such a universalized moral ideal could appear repressive to people unable to live in Kass's ideal family: parents with adopted children, singles, gay or lesbian families, and such.

The authors of this book object to both Kass's description and prescription. It is readily observable that large numbers of persons in our society and other societies are not placed where they can experience family life complete with heterosexual congress and old-fashioned baby-making, yet their dignity is not thereby compromised. Parents with adopted children, for example, create homes with love and companionship just as much as homes with so-called natural families. So also do parents with children born with the help of assisted reproduction, artificial insemination by donor (AID), *in vitro* fertilization (IVF), or even surrogate motherhood. All of these children have dignity, because each is loved as an end of their parent's love. Such families face the same chances of either thriving or facing difficulties in child-rearing that Kass's ideal—so-called natural—family faces.

In addition, the Kass description and prescription leaves out singles prior to entering into their own family ties as well as seniors who have left nuclear family life behind. We have no reason to think their dignity is lost because of this. Gay and lesbian persons, single or in relationships, may or may not elect to bring children into the world; yet their dignity is determined by who they are as persons and not by their family status. We do not believe dignity is contingent on heterosexual congress or old-fashioned family life. Dignity consists of an individual human person being treated as an end rather than a mere means, and this can take place in any of the alternative family structures just mentioned.

Our second critique is aimed at the wisdom of repugnance, introduced in an earlier chapter. We call this the "yuck factor." Even Kass uses the word "yuck."[25] The repugnance some people feel in response to the prospect of cloning is an indication of its moral unacceptability, argues Kass. Such "repugnance is the emotional expression of deep wisdom, beyond reason's power completely to articulate it."[26] Ethics can rely on repugnance as its foundation.

The wisdom-of-repugnance argument relies on the sense of moral abomination, namely, that immediate feeling of repulsion we get when confronted with something strange and apparently ominous. It is immediate, in that it is not inferred from any other ethical category. It sometimes appears as dirt, stain, deformity, obscenity, or profanity. Our response is to just say, "yuck." Kass believes this counts ethically. Although yuck needs to be supplemented with rational moral argument, Kass grants the status of foundational wisdom to the intuitive sense of repugnance.

How should we assess this? Few would allow repugnance to contribute to, let alone count as, an ethical judgment. We know too well that what is repugnant for one pair of eyes is not for another. When we travel to foreign lands and taste for the first time the cuisine of different cultures, Gaymon delights in eating everything. Even if the legs are still squirming in his mouth, he smiles and says "yum." The mere sight of it causes Ted to exclaim,

"Yuck!" Would we want to base our ethics of eating on the wisdom of Ted's repugnance? No.

The forms that repugnance takes are contingent, and can change from culture to culture or over time within a single culture. Only a century ago in North America the public was repulsed at marriages between African Americans and Caucasians. Now such relationships are commonplace. Nazis in Germany in the 1930s propagandized the citizenry in order to create a sense of repugnance against the Jews among them. Such social repugnance served a racist ideology that led to genocide. Virtually no one today would embrace such an ethic. In general, the feeling of revulsion in race relations has undergone an enormous reversal, and so also the values and ethics associated with racial equality. In fact, it took national and international shared reasoning combined with social activism to overcome the former wisdom-of-repugnance argument to establish justice and equality. In short, we do not trust the wisdom-of-repugnance argument as a source for moral standards. We would prefer that our ethical deliberation rely upon foundational reasoning as well as appeal to principled apprehensions of what is good.

Finally, we find insufficient the Kass argument against human embryonic stem cell research. We start by thanking Kass for taking up the questions raised by ethicists who operate within the embryo protection framework. Does life begin at fertilization? Yes. On what grounds? Because God imparts an immortal soul? No. On the grounds that it is biologically observable. "Any honest biologist must . . . view that a human life begins at fertilization." Kass then asks whether the blastocyst equals a human being or not. No, but it is potentially a human person. "I myself would agree that a blastocyst is not, in a *full* sense, a human being—or . . . a person. . . . Yet, at the same time, I must acknowledge that the human blastocyst is (1) human in origin and (2) *potentially* a mature human being, if all goes well."[27] If Kass would restrict himself to the embryo protection framework, we do not see sufficient reason here to either support or oppose therapeutic cloning and stem cell research. More is needed to make a full moral argument against stem cell research.

Now, Kass wants to protect human nature, not embryos per se. At this point, the argument gets vague. He appeals to a connection with the family, especially with procreation. What is distinctively human by nature is what draws Kass's attention. If it has potential for becoming human, it should receive moral respect. "In the blastocyst, even in the zygote, we face a mysterious and awesome power, a power governed by an immanent plan that may produce an indisputably and fully human being. . . . It deserves our respect."[28] Stem cell research allegedly fails to show respect. Kass's destination is the same as that of the pope, even if the route to get there differs.

What does not appear in this train of ethical reasoning is the scientific potential for saving persons from suffering or enhancing their well-being. The only standard here is our given human nature, and somehow in Kass's logic the use of *ex vivo* blastocysts for laboratory research violates human dignity.[29] We simply do not find this to be a persuasive case, either within the human protection framework or when comparing his position with those found within the future wholeness framework.

Having stated these critiques, we want to pause here and recognize the significant value of Kass's contribution to how public ethics is conducted. By reframing the task of bioethics in terms of "the full human meaning of biomedical advance," Kass has to a large extent met his goal of contributing to "a richer and deeper public bioethics." Kass helped create a public space for ethical reflection within which the wisdom of philosophical and theological traditions could be taken seriously and brought to bear on contemporary problems. He helped foster methods of reflection within which the question of what is good and valuable could be posed beyond mere means-ends calculations. He recognized and resisted the demand that science merely cater to social and economic interests wherein ethics might merely serve as the mediator and enforcer of such interests.

In short, where we disagree with Kass is not in his search for a bioethics connected to full human meaning. Rather, we have a different idea of what constitutes full human meaning. While we agree that human life is integrally connected to nature, and also that it is characterized by a striving to overcome limitations, we disagree that the "truly human" is a static balance between these two. We think that what it means to be human is dynamic and changing. We are part of a natural world that is evolving, susceptible to being reshaped, and even, in some cases, made better through scientific innovation. As mentioned in an earlier chapter, human nature includes our destiny as the created cocreator.

Again, this does not mean that we are writing a blank ethical check to be cashed in the name of any new scientific development under any circumstances. Many technical advances do not contribute to human flourishing—Kass is certainly correct on this point! Humans have a long track record of both intended and unintended evils. While recognizing all this, we think that creation is continuing; nature is not fixed. Our human essence is not something already defined, something grounded in our past. As we have already stated: God is not done creating us yet. And, our contribution to creativity within this world—even our bioscientific contribution—is godly in character. To strive to contribute to future wholeness is a key part of what makes us truly human.

A Postscript from the Embryo Protection Framework

Human protection is not the only framework with which Kass is familiar. He can also endorse his colleagues who work within the embryo protection framework.

In 2005 the PCBE published a white paper, *Alternative Sources of Human Pluripotent Stem Cells*. This is the white paper that included William Hurlbut's proposal for ANT (altered Nuclear Transfer). The task of this investigation, recall, was to find sources for pluripotent stem cells that could protect the early embryo from laboratory destruction. If such cells could be gained from alternative sources, then embryos would be spared. In his "Letter of Transmittal to the President of the United States," Kass writes that "science itself might provide a way around this ethical dilemma."[30] That is to say, by accepting the ethical issue as an embryo protectionist would frame it, Kass seeks to provide a scientific solution.

The four proposals analyzed in this report include: (1) deriving pluripotent stem cells from organismically dead embryos, a proposal put forth by D. W. Landry and H. A. Zucker; (2) extracting a single viable blastomere from a living embryo, permitting the remaining blastomeres to develop into a fetus; (3) creating an embryo by genetic manipulation so that it will be unable to develop into a fetus, and then removing viable pluripotent cells, the ANT proposal from William Hurlbut; and (4) employing somatic cell dedifferentiation as a means for creating an artificial embryo for laboratory research. The first three create scientific problems because they risk passing early genetic defects through the stem cell lines to the patient receiving regenerative therapy. The fourth, somatic cell dedifferentiation, offers a tantalizing theoretical possibility and at the time of this writing is possibly on the horizon, but much work remains to be done in order to determine whether it is feasible for therapeutic purposes.

Regardless of the scientific status of these four proposals, what we see here on the part of the PCBE is an attempt to resolve a fundamentally theological issue through scientific means. And, because the issue is formulated within the embryo protection framework, Leon Kass needs to step out of his own human protection framework temporarily in order to accommodate Washington's wishes.

Leon Kass is of Jewish origin, even though his scholarship roots itself in general human nature and not restrictively within the tradition of Jewish theology. In the next chapter, we will turn to bioethics pursued from a distinctively Jewish perspective, and an Islamic perspective as well. When ethicists appeal to the theological roots of these two traditions, they largely find themselves working within the future wholeness framework.

Notes

1. The transcript from the inaugural meeting of the President's Council can be found online at www.bioethics.gov/transcripts/jan02/jan17session1.html.

2. President's Council transcript.

3. See Albert R. Jonsen, *The Birth of Bioethics* (New York: Oxford University Press, 1998), 178.

4. Leon Kass, "Making Babies—The New Biology and the 'Old' Morality," *The Public Interest* 26 (Winter 1972): 18–58.

5. Leon R. Kass, *Life, Liberty and the Defense of Dignity* (San Francisco, CA: Encounter Books, 2002), 2.

6. Kass, *Life*, 22.

7. Kass, *Life*, 9.

8. Kass, *Life*, 160.

9. Kass, *Life*, 18.

10. Kass, *Life*, 85, Kass's italics.

11. Kass, *Life*, 32.

12. Kass, *Life*, 269.

13. Kass, *Life*, 270–71.

14. Kass would not concur with the Reproductive Liberty position taken by someone such as John A. Robertson, *Children of Choices: Freedom and the New Reproductive Technologies* (Princeton, NJ: Princeton University Press, 1994).

15. Robertson, *Children*, 171.

16. President's Council on Bioethics, *Beyond Therapy: Biotechnology and the Pursuit of Happiness* (New York: HarperCollins, Regan Books, 2003), 306.

17. President's Council, *Beyond Therapy*, 287.

18. President's Council, *Beyond Therapy*, 288.

19. President's Council, *Beyond Therapy*, 289.

20. President's Council on Bioethics, *Being Human: Readings From the President's Council on Bioethics* (Washington, DC, 2003); excerpts available online: www.bioethics.gove/bookshelf/reader/chapter1.html#introduction.

21. President's Council, *Beyond Therapy*, 11.

22. President's Council, *Beyond Therapy*, 2, Council's italics.

23. Roger N. Hancock, *Twentieth Century Ethics* (New York: Columbia University Press, 1974), 14.

24. Kass, *Life*, 144.

25. Kass, *Life*, 146.

26. Kass, *Life*, 150.

27. Kass, *Life*, 88.

28. Kass, *Life*, 89.

29. Just what is the dignity violated here? The answer is less than clear. Cynthia Cohen criticizes Kass on this point. "He [Kass] does not explain what it is that gives embodied human life . . . this worthiness and consequently leaves us with only a vague notion of what he means by human dignity." *Renewing the Stuff of Life: Stem Cells, Ethics, and Public Policy* (Oxford, UK: Oxford University Press, 2007), 124.

30. President's Council on Bioethics, *Alternative Sources of Human Pluripotent Stem Cells* (Washington, DC, May 2005), ix–x; www.bioethics.gov.

~

CHAPTER FOURTEEN

Jewish and Muslim Bioethics

As we mentioned in an earlier chapter, our own support for beneficence within the future wholeness framework is attuned to the Jewish commitment to *Tikkun Olam*—the responsibility to join God in repairing and transforming a broken world. It presumes that God's creation is not done yet. It is still on the way. We look to the future rather than the past to discern God's will. And God's will includes creative and redemptive activity yet to come. In short, healing and transforming are godly. The potential for future wholeness will play a decisive role in Jewish ethical thinking.

When it comes to ethical issues surrounding stem cell sources, Jewish and Muslim ethicists are less likely than Christians to have problems with derivation. In neither of these two traditions do we find feet firmly planted in moral concrete regarding conception as the start of individual personhood. Human personhood, as we ordinarily think of it, is not established until at least 40 days after conception according to both traditions. So, the derivation of stem cells from preimplantation blastocysts is not likely to stir opposition from within an embryo protection framework from Jews or Muslims.

Jewish Perspectives on Stem Cell Ethics

During the summer of 2001 on the eve of President George W. Bush's August 9 pronouncement restricting federal funding to existing stem cell lines, Jewish voices could be heard on the stem cell controversy. On June 26, 2001, the Union of Orthodox Jewish Congregations of America issued a statement: "Our

Torah tradition places great value upon human life; we are taught in the opening chapters of Genesis that each human was created in G-d's very image. The potential to save and heal human lives is an integral part of valuing human life from a Jewish perspective. . . . We support . . . federal funding for embryonic stem cell research." A month later, on July 18, 2001, the Religious Action Center of Reform Judaism offered a parallel statement: "Our Jewish tradition reminds us that while only God can create life, God has charged humans with doing everything possible to preserve it. . . . Cutting off funding for medical research that has such tremendous potential benefits . . . is both immoral and unethical." These are clear and forceful endorsements of scientific research on stem cells.

Jewish thinkers place ethical concerns within the larger context of God's creation, complete with God's mandate to pursue healing. "For Judaism, God owns everything, including our bodies," writes bioethicist Elliot N. Dorff. "God lends our bodies to us for this duration of our lives, and we return them to God when we die." What this implies is that "God can and does assert the right to restrict how we use our bodies."[1] This leads directly to the mandate to heal, to moral support for the practice of medicine. "Because God owns our bodies, we are required to help other people escape sickness, injury, and death."[2] Support for clinical medical practice implies, in addition, support for scientific research on behalf of human health and well-being.[3]

Dorff applies these commitments to the stem cell controversy. "The potential of stem cell research for creating organs for transplantation and cures for diseases is, at least in theory, both awesome and hopeful. Indeed, in light of our divine mandate to seek to maintain life and health, one might even contend that from a Jewish perspective we have a *duty* to proceed with that research."[4] It is our duty, based upon a command implicit in God's creation, to pursue stem cell research.

This appeal to duty—to a deontological founding of ethical commitment—is reiterated by Laurie Zoloth. The prospect of healing lifted up by the promise of stem cell research is a call to duty. "We have a duty to heal, and this is expressed in legal and social policy. To turn from the possibility of healing would be an abrogation of an essential duty."[5]

Rabbi Moshe David Tendler, while testifying before the U.S. President's National Bioethics Advisory Commission, said that "mastery of nature for the benefit of those suffering from vital organ failure is an obligation. Human embryonic stem cell research holds that promise."[6]

The pursuit of moral guides within Jewish thinking relies on Torah interpretation. The most frequently appealed to method for Jewish theology and ethics is one of interpreting the Torah through the history of texts that make up the Hebrew and Jewish traditions. Moral laws are derived through application of interpretations, through *halakhah*. Yet, this is more than appeal to tradition as authority. Wisdom is to be found via *halakhah*. Knowledge of human nature and knowledge of God's will expressed in all of nature can be found in this way. The resulting moral laws are not strictly positive—that is, not limited to grounding in the religious legal system. They are also rooted in the universal natural condition of which Jewish reflection is a cultural expression. Wisdom asks us to consider their universality. For this reason, one can legitimately move from the tradition-specific character of Jewish *halakhic* reasoning in the direction of public policy. If ethics is grounded in universal human nature, then a pluralistic society can benefit from such religious wisdom without dismissing it as the vested interest of a single sect.

Because of this connection between God's will, God's creation, and our nature, could the question of playing God arise? Are Jewish ethicists likely to worry that laboratory scientists might profane something sacred in nature and precipitate retaliation in the form of biological disaster? Do they prohibit playing God in medicine? Ordinarily, the risk of playing God is not a large factor in Jewish ethical deliberation. Laurie Zoloth writes, "whereas moderns are worried lest we 'play God,' the rabbis were concerned that we act *more* like God might in many ethical and social-political arenas, as in helping the poor, creating justice, and healing the sick."[7] Yet, appeal to the secular commandment to avoid playing God may still arise. Elliot Dorff asks, "How do we determine when we are using genetic engineering appropriately to aid God in ongoing, divine acts of cure and creation and when, on the other hand, we are usurping the proper prerogatives of God to determine the nature of creation? More bluntly, when do we cease to act as the servants of God and pretend instead to be God?"[8]

Jewish Theology and Embryo Protection

As we suggested above, Jewish ethicists will not have the difficulties with hES cell derivation that we find among Christian embryo protectionists. The Jewish view of the early embryo differs markedly from that of the Vatican in two ways. First, human personhood whether defined as quickening, dignity, or ensoulment does not register until 40 days after conception. The Mishna (Nidda 30a) states that a miscarriage prior to 40 days does not cause *tumat leida* (impurity). This leads Daniel Eisenberg to say that "a fetus prior to forty days gestation is not considered to be an actual person and we might extrapolate that destruction of such a fetus is not forbidden by Jewish law."[9]

For some Jews, birth is the threshold for identifying individual human life and for establishing moral protectability. This has permitted many Jews to support the pro-choice side of the abortion debate. This simply carries over into the stem cell debate.

Second, within the context of Jewish thinking, the concept of abortion applies to the removal of a fetus from a woman's body. It does not apply to what happens in an *in vitro* fertilization (IVF) clinic or a laboratory. One can abort a fetus *in vivo*, but the destruction of a blastocyst *ex vivo* does not fit the definition of an abortion. The category of abortion applies only when an embryo is in a woman's uterus; for the embryo has the potential to develop and become complete only when living in her uterus. "The early *in vitro* embryo has a somewhat lower status than *in utero* fetus," writes Aaron Mackler. "The former has not reached the stage of individualization, and because of its location it is not on a natural trajectory of development."[10] Because the early embryo outside a woman's body is not viable, it is not a potential person. Therefore, it does not have morally protectable status. If one is to prohibit abortion, then it would apply to the removal of an embryo or fetus from the woman's body. Such removal does not take place in stem cell research.

In Jewish bioethics, the distinction between *in vivo* and *ex vivo* is significant. "Genetic materials outside the uterus have no legal status in Jewish law," says Elliot Dorff, "for they are not even a part of a human being until implanted in a woman's womb and even then, during the first 40 days of gestation, their status is 'as if they were water.'"[11] Regardless of when individuation and moral protectability occurs within the womb, these do not apply to *ex vivo* laboratory research. To be a potential person, one must be located where that potentiality can become actualized. And only the mother's womb can provide that. This Jewish position stands in sharp contrast to the Vatican position, because the Vatican applies the same moral judgment to both *in vivo* and *ex vivo* embryos.

Even so, it is possible for a Jewish thinker to find sympathy for the Vatican position. In our discussions within the Geron Ethics Advisory Board (EAB), our Jewish scholar, Laurie Zoloth, followed the *halakhic* method of interpretation. She was enthusiastic about the potential of stem cell research to deliver on promises of a quantum leap forward in the quality of human health. Like her Jewish colleagues, she gave high priority to the ministry of healing. Yet, she found herself repeatedly defending the conservative Christian view on embryo protection. At times she articulated the Vatican position in a manner that would have made Pope Benedict XVI proud. What Laurie was giving voice to is the coherence of the embryo protection position, once one grants certain premises. Jewish bioethics is grounded in a theology that does not grant the same premises. As a fair scholar, Laurie could defend alternative sides in the debate.

Laurie Zoloth gave voice to what all of us on the Geron EAB believed: "Moral vision has to precede research."[12] If science is to serve, the research must locate itself within a moral vision that will lead to healing and flourishing for the human race.

Bioethics in Islam

Like Jews and Christians, Muslims honor and support the science of medicine because Islam is a religion that seeks the health and well-being of each human being. The prospects of regenerative medicine have attracted considerable attention in the world of Islamic science.

We forecast that Muslims will increasingly support human embryonic stem cell research. This is already the case in America. In a survey conducted by the Islamic Institute in Washington in 2001, 629 individuals were polled. Of these, 394 (or 62%) stated their overall support for research on human embryos. A total of 457 (or 73%) stated that it is acceptable to use embryos that have already been donated from *in vitro* fertilization; and 383 (or 61%) stated their support for using embryos to be donated in the future. A total of 312 (or 49%) felt it is acceptable to produce embryos specifically for stem cell research purposes. The Islamic Institute itself strongly supports transferring excess embryos from freezers into laboratories. "It is a societal obligation to perform research on these extra embryos instead of discarding them."[13]

Stem cell research centers are springing up in Egypt, Iran, and other Islamic states. Iranian scientists developed human embryonic stem (hES) cell lines in 2003 with the approval of religious leader Ayatollah Seyed Ali Khamenei. The die is not yet cast, but we expect to see developing theological support for stem cell research.

As we talk with Islamic scholars around the world, we hear frequently a sense of urgency that more Muslims need to become trained in bioethics to provide necessary leadership in this fast-moving world of bioscience. When it comes to knotty problems, such as the stem cell debate, Islamic theology is underprepared. Yet, public policy decisions need to be made in Islamic countries, and the conversation is robustly proceeding.

Like Jews, Muslims develop their moral principles through a process of interpretation. When what is required for interpretation is an ancient text, such as the Qur'an, this takes time. Do we have sufficient time for each scientific controversy that breaks out?

Recall for a moment the international controversy over reproductive cloning in 1997. At that time, the U.S. National Bioethics Advisory Commission invited religious leaders to Washington to provide testimony for what
might become government policy. Aziz Sachedina, a Muslim bioethicist at
the University of Virginia, testified that the Qur'an and subsequent Islamic
tradition had not provided background or principles that anticipated modern
biological knowledge about the embryo or genetic inheritance.[14] Definitive
moral guidance must await a process of interpretation that will involve application of past tradition to present circumstances. In Islam one appeals not
to nature plus scripture, but to scripture alone, to the Qur'an. When Muslims
get to the task of consulting the Qur'an, they will be committed to a literal
reading. And they anticipate that what they will find in this holy book will
be God's very word regarding scientific truths.

Sachedina eventually formulated an opinion on reproductive cloning,
however. "Unanimity has now emerged among Muslim scholars . . . the idea
of human cloning has been viewed negatively and almost . . . 'Satanic.'"[15] In
Islamic contexts, we are seeing the rise of opposition to reproductive cloning
alongside support for stem cell research. Sachedina himself writes elsewhere,
"Hence, in Islam, research on stem cells made possible by biotechnical intervention in the early stages of life is regarded as an act of faith in the ultimate
will of God as the Giver of all life, as long as such an intervention is under-
taken with the purpose of improving human health."[16]

The Qur'an and the Soul

The Qur'an is the preeminent source for moral direction in Islam, though it
does not stand alone. The holy book is accompanied by its tradition of elaboration in the Sunna (meaning "trodden path"). To these two, the Qur'an and
the Sunna, are added two other sources, consensus (ijma) of the early Muslim
community plus the principle of analogy (qiyas).[17] Analogy is a method of
reasoning from data supplied by the Qur'an and the Sunna in which the
unknown is approached via analogy to what is known.

As ethicists and jurists within Islam confront new and unprecedented
scientific challenges to human self-understanding, appeals to the Qur'an and
ancient tradition begin their analysis. To classical theological vocabulary
only carefully filtered additions of contemporary scientific terminology are
added. This means issues formulated by the contemporary scientific situation
can be addressed only indirectly rather than directly. This means the method
will necessarily be one of analogy (qiyas). Analogs to past juridical deliberations will be retrieved and contemporary applications sought.

Included within the tradition that makes up Islamic thought are philosophical debates regarding such subjects as the human soul, especially the relationship of the soul to the body. We predict that ancient debates on the soul will influence contemporary jurists (*fuquaha*) as they pursue the science of jurisprudence (*usul al-fiqh*). Might we expect a twenty-first-century extension of a fissure that opened up in the eleventh century, namely, the split between Ibn Sina (980–1037) and Abu Hamid al-Ghazali (1058–1111)?

Let us take a moment to contrast these two versions of the soul. Ibn Sina, whom the Christian theologian Thomas Aquinas cites as Avicenna, held a view that comes close to what Plato believed. Here, the soul is incorruptible and does not die with the body. Consciousness can continue to exist in a disembodied state. Attachments to the body that involve temperaments are accidental, not essential, to the soul. Our body is not the form of our soul; nor does the soul imprint itself onto the composite parts of the body. The soul is not intrinsically dependent on the body. Rather, the soul's fundamental relationships are with eternal principles that escape what we experience as change or corruption. The result of this substance dualism is a form of everyday naturalism. By "naturalism" we mean this: Whatever happens to the body is exhaustively explained by its place in the physical causal nexus. The body is not tied to what is eternal. We do not need heavenly explanations for earthly happenings. The body's natural nexus of activity drops into near insignificance compared to the soul's out-of-body destiny.[18]

Whereas Ibn Sina is an Islamic disciple of Plato, his opponent, al-Ghazali, is more sympathetic to Aristotle. Ghazali says that whenever we take an action, that action is initiated by our soul. This means, contrary to Ibn Sina, that the soul cannot avoid an inextricable attachment to bodily movement. Our physical causal nexus and the social nexus include the soul's activity. A continuity exists between the spiritual substance of the soul and the physical substance of the material world, though they are not identical. Ghazali describes the body as the camel. The soul mounts the bodily camel to ride toward God. Without the body, the soul cannot reach its destination. Whereas stricter dualists such as Ibn Sina could think of life beyond death in terms of a disembodied soul; Ghazali holds out for a bodily or corporeal resurrection.[19]

Now, this is an ancient dispute. Yet, these ancient disputes continue to live in a tradition that needs to make sense out of revolutions in science. Here is an example. Ebrahim Moosa demonstrates that these two views of the soul influence two contrasting positions on the question of organ transplantation from a person who is brain dead. He compares two *fatwas* or nonbinding juridical opinions, one from Pakistan and one from Egypt. In the case of the Pakistani *fatwa*, organ transplantation violates human dignity (*karam wa hurma*). The person declared brain dead is the one whose dignity is being protected here; dignity is preserved by not dismembering the corpse. Repelling harm to the body—even the dead body—takes precedence over potential medical benefit to someone else. Moosa concludes that the position of Ghazali is being defended here.

In contrast, the Egyptian *fatwa* permits organ transplantation. In this argument, it is assumed that the soul's presence is the source of animation, and hence it is tied to brain function. Because a brain-dead person is no longer animated by the soul, and because the soul has been released to eternity, what remains is a body subject solely to the physical causal nexus. The dead body is solely natural, with no supernatural component remaining. Organ transplantation is permissible. Note the strict dualist assumption. "Once it could be argued that the locus of the soul is the brain and that consciousness is an indicator of brain function, brain death can easily be justified. Those jurists who opposed brain death and organ transplantation used the same texts and sources as their fellow jurists but arrived at an opposing and differing position. Their emphasis was on the social imagery of the body as inviolable in its dignity."[20]

What we learn from this is that, like all other traditions, interpreting holy texts and traditions inevitably involves a diversity of perspectives, contexts, points of view, and opinions. What theologians call the "hermeneutical process" will continue in every major religious tradition founded in classical times. As Islam confronts the modern world of science, this ought to be expected.

As the Islamic tradition grapples with ethical issues surrounding stem cell science, what might we forecast? Regarding the future wholeness framework and the potential for improving human health and flourishing, we could anticipate that Ibn Sina's sympathizers could celebrate the medical advances but not necessarily see implications for one's eternal destiny. The health of the soul will not be influenced by the improved health of the body, even improved health of brain functioning. Sympathizers with Ghazali, on the other hand, could very well embrace stem cell therapy as an enhancement to the body that enables a fuller soul function in this physical world, and perhaps even an incentive to speed the camel in its pilgrimage toward Allah.

Stem Cell Derivation from Early Embryos in Islam

With regard to stem cell derivation, Muslims are not likely to block procurement, although articulated opinions on this matter are less than uniform.

In some sections of the Qur'an we find a threshold dated at 40 days after conception, elsewhere ensoulment at 120 days (4 lunar months plus 10 days after conception, or the first trimester). Most Sunni and some Shi'ite theologians distinguish two stages of pregnancy. The early stage is pre-ensoulment. It is biological, but not yet spiritual or moral. After the fourth month or 120 days, the time of ensouled quickening, the biological person becomes a moral person.

Embryology in the Qur'an looks like this: "We created man from the quintessence of mud. Thereafter, We cause him to remain as a drop of sperm in a firm lodging (the womb). Thereafter We fashioned the sperm into something that clings (*Alakah*), which We fashioned into a chewed lump (*Modgha*). The chewed lump is fashioned into bones which are then covered with flesh. . . . The creation of each one of you is collected in forty days. . . . Then the soul (*Rooh*) is breathed into him. . . . Blessed is God, the best of artisans" (Sura 23/12ff.). What we see here is a developmentalist interpretation. Within the mother's body the embryo passes through stages of development until flesh covers the bones. Then, God blows the soul (*Rooh*) into the physically developed embryo.

Yet, this event at 40 days does not suffice for providing moral protectability. That waits for 120 days, 17 weeks and 1 day. Quickening can be felt by the mother at 15 weeks at the earliest. "So long as the foetus has not reached the 120 days, it is permissible, in the view of the jurists, to perform abortion if indicated medically."[21] There is no resemblance here to contemporary Roman Catholic commitment to ensoulment accompanied by dignity at conception. "The embryo is not considered a person and the use of it for stem cell research does not violate Islamic law."[22]

This variety leads Sachedina to conclude that no impediment stands in the way of deriving hES cells from early embryos in the laboratory. "The silence of the Qur'an over a criterion for moral status of the fetus allows the jurists to make a distinction between a biological and moral person, placing the latter stage after, at least, the first trimester in pregnancy . . . [this will] allow their use for stem cell research."[23]

Genetic Continuity in Islam

Curiously, an additional argument is being raised within Islamic circles supporting donation of excess fertilized ova for stem cell research. Here's how the argument works. Inheritance is extremely important in cultures defined by Islamic tradition. Inheritance is dependent upon blood lines; so genetics is an area of science put to use in determining just who is eligible to inherit family property. Clarity in this regard is paramount.

Muslims who take advantage of reproductive services such as IVF worry about the excess fertilized ova in frozen storage. Might a mistake occur? Might one or more of these frozen zygotes accidentally get planted in another woman? Might there be a possibility—even if remote—that one family's genes might appear in the genome of a stranger? Could that person eventually make a claim on inheritance? It is a nightmarish thought.

The worry can be eliminated if all frozen embryos are eliminated. Muslim families are now standing up and offering their excess embryos for laboratory use, because it guarantees that genes with potential inheritance claims will not get out. The result is that laboratories will find a source for research materials.

The Islamic Organization for Medical Sciences (IOMS)

In summary, Islamic bioethicists take a stand against reproductive cloning while favoring stem cell research and genetic therapies more generally. Meeting at the World Health Organization offices in Cairo, Egypt, in early February 2006, the IOMS centered in Kuwait hosted a conference of Muslim, Jewish, and Christian bioethicists. The aim was to produce a set of principles common to the three Abrahamic traditions that could guide laboratory researchers and medical practitioners through the genetic thickets. Principle 10 of this "Declaration of Principles" points the direction taken. "It is permissible to utilize genetic engineering to diagnose disease or to cure or alleviate human suffering. It is permissible to insert a gene from one nonhuman being into another nonhuman to obtain large quantities of the secretions of such gene for use as a treatment for some diseases. The State should provide such treatment for all citizens in the same manner as it guarantees access to other forms of health care."

During the first week of November 2007, the World Health Organization in Cairo, Egypt, hosted a follow-up conference on "The Dilemma of the Stem Cell: Research, Future, and Ethical Challenges." Again, the organizing group was the IOMS with cooperation from OIC, UNESCO, ISESCO, and CIOMS. It brought together 80 researchers from Egypt, Kuwait, Saudi Arabia, United Arab Emirates, Jordan, Syria, Morocco, Algeria, Tunisia, the United Kingdom, and the United States. In addition to Muslim scientists, theologians, and jurists, Christian stem cell researchers and bioethicists were invited to deal with the seemingly intransigent dilemma presented by the embryo protectionists in the Christian West.

Some sympathy for the Vatican and related embryo protection positions could be found among Islamic scholars. However, the preponderance of Islamic opinion fell on the side of saying "yes" to stem cell research. No decisive theological impediments to regenerative medicine employing hES cells could be raised. The first proposed recommendation from the international seminar reads: "It is not religiously impermissible to conduct research on stem cells to produce body tissues with an eye to using them in the treatment of some diseases on condition that these cells are obtained from religiously permissible sources." The second reads: "IVF surplus fertilized eggs are by no means impermissible and enjoy no sanctity before their implantation."[24] Reproductive cloning was again rejected, but therapeutic cloning to establish stem cell lines that would overcome immune rejection was recommended. Use of umbilical cord blood and other sources of adult stem cells was encouraged. Chimerism to advance our knowledge was encouraged; but the recommendations included maximal caution in xenotransplantation to prevent transfer of animal diseases to the human patient and to avoid humanizing the mentality of animals. As of this writing the IOMS plans to join with the World Health Organization to construct a set of guidelines: "The Islamic code of the good practices for the use of human stem cells."[25]

Conclusion

Are the cells within a blastocyst sacred for Jewish and Muslim adherents? No. The *ex vivo* embryo is certainly valued and respected as a life-giving source, to be sure. But, prior to 40 days following conception and only in a mother's womb do we find a human person with either potential or actual dignity.

If we were to place Jewish and Muslim ethical deliberation into one or another of our frameworks, we would find that with greater or lesser intensity the following: Regarding the question of embryo protection, neither Jewish nor Islamic theology or ethics would support the Vatican position that associates conception with ensoulment, dignity, and *ex vivo* restrictions. The *ex vivo* preimplantation blastocyst used by stem cell researchers does not warrant the status of an individual human being, or a potential human being, requiring protection from research.

Regarding the future wholeness framework, both Jews and Muslims emphasize that we have a moral obligation to serve the health and well-being of human persons. This more than justifies society's support for stem cell research. It requires it.

Notes

1. Elliot N. Dorff, *Matters of Life and Death: A Jewish Approach to Modern Medical Ethics* (Philadelphia, PA: Jewish Publication Society, 1998), 15.

2. Dorff, *Matters of Life and Death*, 26.

3. Jewish voices have been heard in support of stem cell research. Jewish theology emphasizes the divine mandate to steward medical science in the service of human welfare, and this applies positively to stem cell research. See www.ou.org/public/statements/2001/nate34.htm and http://uahc.org/cgi-bin/resodisp.pl?file=fetaltissue&year=1993o.

4. Elliott N. Dorff, "Stem Cell Research—A Jewish Perspective," in *The Human Embryonic Stem Cell Debate: Science, Ethics, and Public Policy*, ed. Suzanne Holland, Karen Labacqz, and Laurie Zoloth (Cambridge, MA: MIT Press, 2001), 92.

5. Laurie Zoloth, "Immortal Cells, Moral Selves," *Handbook on Stem Cells*, 2 volumes, ed. Robert Lanza (Amsterdam: Elsevier Academic Press, 2004), II: 753.

6. Moshe Tendler, "Stem Cell Research and Therapy: A Judeo-Biblical Perspective," *Ethical Issues in Human Stem Cell Research*, vol. 3: *Religious Perspectives* (September 1999), at NBAC website: http://bioethics.gov/pubs.html.

7. Laurie Zoloth, "The Ethics of the Eighth Day: Jewish Bioethics and Research on Human Embryonic Stem Cells," in *The Human Embryonic Stem Cell Debate*, 96.

8. Dorff, *Matters of Life and Death*, 162.

9. Daniel Eisenberg, M.D., "Stem Cell Research in Jewish Law," *Jewish Virtual Library*, www.jewishvirtuallibrary.org/jsource/Judaism/stemcell.html.

10. Aaron L. Mackler, *Introduction to Jewish and Catholic Bioethics: A Comparative Analysis* (Washington, DC: Georgetown University Press, 2003), 168.

11. Elliot N. Dorff, "Testimony," in National Bioethics Advisory Commission, *Ethical Issues in Human Stem Cell Research*, vol. 1: *Report and Recommendations of the National Bioethics Advisory Commission* (Rockville, MD: National Bioethics Advisory Commission, 1999), 50.

12. Laurie Zoloth, "Jordan's Banks: A View From the First Years of Human Embryonic Stem Cell Research," in *The Human Embryonic Stem Cell Debate* (Cambridge, MA: MIT Press, 2002), 236.

13. Islamic Institute's ii issues, www.islamicinstitute.org/i3-stemcell.pdf#search='Muslim%20Stem%20Cell.

14. Abdulaziz Sachedina, "Testimony to the National Bioethics Advisory Commission," March 14, 1997 (Washington, DC: Eberline Reporting Service), 56–64.

15. Abdulaziz Sachedina, "Human Clones: An Islamic View," in *The Human Cloning Debate*, ed. Glenn McGee (Berkeley, CA: Berkeley Hills Books, 1998), 240–41.

16. Cited by Michael Weckerly, pfaith.org/islam.htm.

17. Abdulaziz Sachedina, *Islamic Biomedical Ethics: Issues and Resources* (Islamabad, Pakistan: Comstech, 2003), 14.

18. Ibn Sina, *Avicennas Psychology: An English Translation of Kitab al-Najat*, book II, chapter IV, trans. Fazlur Rahman (London: Oxford University Press, 1952).

19. Abu Hamid al-Ghazali, "al-Munquidh min al-Dalal," in Majmu'a Rasa'il al-Imam al-Ghazali, ed. Ahmad Shams al-Din (Beirut: Dar al-Ktuub al-'Ilmiyyah, 1409/1988).

20. Ebrahim Moosa, "Interface of Science and Jurisprudence: Dissonant Gazes at the Body in Modern Muslim Ethics," in God, Life, and the Cosmos: Christian and Islamic Perspectives, ed. Ted Peters, Muzaffar Iqbal, and Syed Nomanul Haq (Aldershot, UK: Ashgate, 2002), 344.

21. Mohammed Ali Albar, Human Development as Revealed in the Holy Quran and Hadith (Dammam and Riyadh: Saudi Publishing and Distributing House, 2002), 137.

22. Michael Weckerly, "The Islamic View on Stem Cell Research," at pfaith.org/islam.htm.

23. Abduliziz Sachedina, "Islamic Perspectives on Stem Cell Research," unpublished paper, 2005.

24. "Recommendations of Stem Cells Seminar," sent by private communication to participants on February 8, 2008.

25. Islamic Organization for Medical Sciences, www.islamset.com/ioms/index.html.

~

CHAPTER FIFTEEN

The Terror of the Chimera

When the National Academies of Sciences (NAS) published its guidelines for stem cell research in the spring of 2005, it included a curious recommendation: "no animal into which hES cells have been introduced at any stage of development should be allowed to breed."[1] The National Council of Churches of Christ in the USA (NCC) made an even stronger statement. They "oppose the creation of chimeras, or any experimentation that might lead to an intermediary human/animal species."[2] What is going on here?

What is commonplace is that researchers frequently find the need to transfer human stem cells into animal models to watch what happens. Such experimental animals are typically referred to as *chimeras*. So far, so good. What the National Academies is recommending is that these animals not be permitted to breed; the NCC appears to recommend that they not be made at all. What is the fear this attempts to relieve? Does this indicate anxiety on the part of someone, somewhere? If so, why?

Perhaps the National Academies had become intimidated by a March 17, 2005, bill introduced in Washington (S.659) known as the "Human Chimera Prohibition Act of 2005." Senator Brownback, author of this proposal, holds "respect for human dignity and the integrity of the human species." This seems to imply the need for a prohibition against creating embryos that mix species. He wants to prevent chimeras that would "blur the lines between human and animal, male and female, parent and child, and one individual and another individual." Just in case this should become law, perhaps the National Academy was reserving the right to create chimeric cells and embryos but lowering the level of offense by preventing breeding. Yet, we ask, why would such an issue arise?

The public debate over chimerism takes place within the human protection framework. It arises out of a fear that our scientists will so change the natural world we have inherited that moral confusion will result.[3]

The Chimera Myth

One of the fascinating things about modern science is that it keeps the ancient myths alive. Our culture has seen how the myth of Prometheus returns as the mad scientist in stories such as *Frankenstein* and *Jurassic Park*, and how our culture is alive with the commandment to avoid "playing God." So, what about the chimera? Might something mythical be returning here?

In book 6 of the *Iliad*, Homer tells an adventure story of Bellerophon that introduces the chimera. Homer tells us that Bellerophon had "surpassing comeliness and beauty." In modern jargon, he was buff. The buff Bellerophon had already distinguished himself as a brave warrior before visiting Proetus, the king of Ephyre in Argolis, the predecessor of the biblical Corinth. The hospitable king invited the attractive visitor home to dinner. While dining, the king's wife, Antea, Homer tells us, began to lust after the manly guest. When she had a chance, she invited Bellerophon for a secret sexual rendezvous.

Personal honor now becomes decisive. Bellerophon refused the seductress. By doing so he showed respect to himself and to his host, the king. But, as we know, "hell hath no fury like a woman scorned." This was true here. Antea approached her husband and described the reverse of what had actually happened. She complained that Bellerophon had tried to seduce her and that she was the one who refused. "Kill Bellerophon," she ordered her husband, the king.

But Proetus was also a man of honor. He recalled a maxim of Zeus: Do not harm someone whom you have hosted at table. So, rather than deliver death to Bellerophon himself, he plotted that his guest would die at someone else's hand.[4]

Proetus's plot included sending Bellerophon to King Iobates of Lycia with a sealed letter in hand. Iobates just happened to be Proetus's father-in-law, the father of Antea. The sealed letter included a secret message from Proetus that the Lycian king should find a way to put Bellerophon to death, as appropriate revenge.

Upon Bellerophon's arrival in Lycia, the king rolled out the red carpet and hosted the visiting warrior lavishly for nine days. Then, he asked for the letter. Once he had opened the seal and read it, he found himself in the same honor dilemma as Proetus. Iobates could not slay his visitor with his own hand. He too needed to develop a plot that would lead to a natural death, so to speak.[5]

It is at this point that the chimera enters the story. The assignment King Iobates gave to the unsuspecting Bellerophon was to slay the chimera (chimaera). The chimera was of divine origin, a goddess, but certainly not beloved by the Olympian gods. In this single animal we find combined the head of a lion, the tail of a serpent, and the body of a goat. That's what Homer says. Hesiod says that the chimera also included a goat's head. What they agreed on is this: the angry animal breathed fire, a ferocious fire. It was a single creature with the power of three beasts. The chimera was also called *amaimaketos*, meaning "raging" and "invincible." Surely Bellerophon would not return alive.

Defeat would not be Bellerophon's destiny, however. The clever warrior climbed aback his faithful winged horse, Pegasus. He flew to the mountain ledge where the grouchy and aggressive beast maintained her lair. He guided the flying steed above the range of the spitting fire. With archer's accuracy he shot arrows from his bow that found their mark. Bellerophon returned to Iobates and Proetus as the hero, not the slain.

Now, why is this ancient Greek myth important for us today. Why do we say that it "lives"? Because myths such as this are classics; they work silently within our language and culture, forming our minds and framing our understanding. The late German philosopher Hans-Georg Gadamer gave us the difficult-to-pronounce concept of *Wirkungsgeschichte*, translated poorly as "effective history," to show how classics continue to have a history of effects.[6] Greek myths are classics. Even today in our modern secular culture, we still think out of these myths so automatically that we hardly even recognize it. In Western civilization, the word "chimera" can be used in biology for any hybrid plant or animal. In literature we may also use "chimera" metaphorically to identify an imaginary fear. *A chimera is an imaginary fear*. Could the ancient Greek story still have some semiconscious effect on the way we think? Could this long-dead beast continue to terrorize us today?

What Is a Stem Cell Chimera?

What is a chimera in today's biology? Most people work with a broad and inclusive definition. Jason Scott Robert and Françoise Baylis offer a working definition as wide as the mouth of the Mississippi: A *chimera* is "a single biological entity that is composed of a mixing of materials from two or more different organisms."[7] This is a very broad definition, because it does not distinguish what "materials" might get mixed. Could it be tissue? Could it be individual cells? Could it be chromosomes within cells? Could it be genes within the chromosomes? All are swept out to ethical sea by this one current.

Here we plan to narrow the definition to a workable scale: A chimera is a single living organism with two or more genomes. With this definition, we observe that chimeras come in multiple colors and shapes. Once we work with a broad but simple definition—a chimera is a single organism with two or more genomes[8]—we can identify a number of examples: Human to Human (HH) chimerism (surgically placing in a patient pluripotent stem cells derived from someone else); Human to NonHuman (HNH) chimerism (placing human DNA in a mouse oocyte); and perhaps even NonHuman to Human (NHH) chimerism (human stem cells grown in an animal host and then implanted in a human patient) (see table 15.1). The introduction of donated stem cells into the patient will make him or her into a chimera, one person with two genomes.

Some chimeras occur naturally. Human or HH chimeras are spontaneously created during the fertilization process. Perhaps two eggs will become fertilized, then fuse together to become a single embryo. In the case of a double fertilization and fusion a person can be born with two genomes. Such a person is a chimera. Most chimeric people grow old and die and never realize their double genetic identity. The Brownback proposal would have them living out their lives in jail.

In addition to natural chimeras, we can make them in the laboratory or hospital. Leukemia patients who receive blood stem cell transplants become chimeric, now carrying the genetic codes of two persons. In an attempt to establish histocompatibility for future organ tissue transplants, stem cell researchers are experimenting with creating chimeras. The theory is this: First, doctors would inject donated hematopoietic stem cells into the bone marrow. Once these have traveled throughout the body and lodged in the organs, doctors would surgically implant the stem cell–produced organ tissue, derived from the same donated source as the hematopoietic cells. The hope is that the previously integrated blood stem cells would welcome the surgically

Table 15.1. Three Kinds of Genetic Chimeras

	HH	HNH	NHH
H	Human to Human	Human to NonHuman	NonHuman to Human

implanted tissue as their own without rejection. If this works, HH chimerism will become routine in stem cell therapy.

Now, these are examples of chimerism within the human species. We are talking about a single person with two human genomes. To date, this seems to have raised no ethical ire, even though the Brownback proposal literally interpreted would render such chimeric persons illegal.

What happens when the genomes come from two different species, one human and the other nonhuman? Does NHH or HNH chimerism change things, ethically speaking?

We already have considerable experience transferring tissue from animals into human beings. Pig heart valves are now pumping healthy blood for countless human heart patients. Yes, we need to be cautious about passing diseases endemic to pigs into humans. But this is a pragmatic problem, not an ethical one.

What about going the other way? What about transferring human tissue into animals? The same reasoning applies. Sheep with implanted human genes now make vaccines for human diabetes patients. In research, the SCID-hu mouse mentioned earlier in this book has provided immense value to researchers studying immune system dysfunction and HIV. It has "not raised any outcry," comments Stanford University Law School professor, Henry Greely. "Chimeras that are produced 'naturally' seem to raise few concerns. Many 'unnatural' chimeras are also uncontroversial. Chimeras made by moving nonhuman parts into human beings would raise concerns when they are significant enough to cast doubt on the humanity of the recipient. . . . The fact that something is or isn't a chimera does not in itself raise ethical concerns."[9]

Just where does the concern for the ethicist arise? It arises twice in HNH chimeras: first, when the insertion of hES cells into a nonhuman brain risks development of human thinking capacity, and, second, when insertion of hES cells into a nonhuman animal risks becoming part of the germ line leading to chimeric progeny. In 2000 Irv Weissman and his colleagues at Stanford inserted human fetal neural stem cells into the brains of newborn mice.[10] This study sought to follow pathways that hES cells might eventually take in humans during neural development. Did this mouse brain take on human capacities? No, say the researchers. The chemical signals from the host mouse brain controlled the development of the hES cells. The mice remained mice, they say. Yet, we might ask: Are they certain the mice remained mice? Or, what if it had turned out differently? What if the mice had become humanized by such an experiment? This is where the ethical issue of chimerism arises.

What we see here is concern for just what is distinctively human. To be human is to be a subject of ethical deliberation. So, when in the context of contemporary biological discussion the topic of chimerism comes up, watch out for the human element.

The Morality of Species Mixing

Let us refine the issue by posing the following question: If we take human stem cells and place them in the brain of another animal as a host, and if our stem cells affect not just the body but also the mind of that animal, then what? A number of subquestions are embodied in this bigger one. First, do any ethical issues arise from mixing cells from different species? Second, does the effect on the animal's mind make the animal closer to human? Third, does the effect on the animal's mind add something that leads us to cross a moral threshold?

Recently a team of scientists and ethicists writing for *Science* magazine studied the moral issues surrounding transplantation of human neural stem cells into the brains of nonhuman primates (H-NHP). Here is how they posed the questions. "If human neural stem cells were implanted into the brains of other primates what might this do to the mind of the recipient?" And, "could we change the capacities of the engrafted animal in a way that leads us to reexamine its moral status?"[11] These are fascinating questions.

Here is a version of the question that terrorizes some of us: Should we mix the genes of two different species? The aforementioned *Science* team is not terrorized, by any means. "We unanimously rejected ethical objections grounded on unnaturalness or crossing species boundaries." Two reasons are given for making this question such an easy one to answer. First, such a thing as a boundary between species does not exist, scientifically speaking. Oh yes, we have inherited from the era of Darwin and earlier the notion that a species can be defined as a group of organisms that can reproduce. If you cannot make babies together, then you belong to another species. But, new knowledge regarding the consistency of DNA shows the genetic continuity of all living things; such thick lines of separation are fading away. "The notion that there are fixed species boundaries is not well supported in science or philosophy. Moreover, human-nonhuman chimerism has already occurred through xenografting." Chemically, all life belongs together. DNA continuity is what makes mixing genes easy to do.

Some might say: Let's not mix species, because it's not natural! The *Science* team dismisses this objection too. To object to anything a scientist does as "unnatural" does not in itself provide a moral warrant. As we saw in our discussion of the second moral framework, the human protection framework, it is a fallacy to move from what we *describe* in nature to what we ought to *prescribe* scientists to do. This team flatly avoids giving credibility to those who commit this fallacy.

Theological and Secular Arguments against Mixing Species

Despite the ease with which the *Science* report dismisses the problems associated with chimerism, many are still uneasy with the prospect. Some religious bioethicists want to prevent the mixing of species and, thereby, shut down this form of hES research. For example, in its *Prospects for Xenotransplantation* [xenotransplantation generally refers to NHH or HNH chimeras], the Pontifical Academy for Life tries to protect human identity from the threat of chimerism. "The implantation of a foreign organ into a human body finds an ethical limit in the degree of change that it may entail in the identity of the person who receives it."[12] This might be interpreted as qualified rejection of chimerism, not a categorical proscription.

Might we find religious objections anywhere else? Let us speculate about the theological community: Who else might object to species mixing and why? Who might take umbrage at this team's decision? Someone who wants to draw a thick line of demarcation between species and who wants to put up a "no trespassing" sign.

Could this apply to the scientific creationists? This would seem appropriate. The key word in their argument is "kind." This word comes from the creation story in the Bible, in which God says "let the earth bring forth living creatures of every kind" (Genesis 1:24). Today's creationists translate "kind" into "species." So, Duane Gish of the Institute for Creation Research near San Diego can write, "each kind was created with sufficient genetic potential, or gene pool, to give rise to all of the varieties within that kind that have existed in the past and those that are yet in existence today."[13] All the dogs from Chihuahua to Great Dane belong to the single species, dog. All human beings regardless of race or ethnicity belong to the same species, human. These are two different kinds. "No matter what combinations may occur . . . the human kind always remains human; and the dog kind never ceases to be dog kind."[14] Gish's colleague, Henry Morris, writes, "each system and each organism were created specifically the way God designed them to be, and He intended each to retain its own character."[15]

We should note, however, that the context of this defense of each species is the battle over Darwinian evolutionary theory, wherein one species evolves into a new and distinct species. The creationists deny the existence of macroevolution. So we might ask: Would their defense of fixed species each with their own character lead to a moral judgment against deliberate creation of chimeras in medical research? To date, creationists have not placed chimerism on their lists of things to fight about. We have to make the argument for them.

How far do we need to run our imaginations to make a theological case against species mixing? What this indicates is that mainline biblical theology simply has no undisputed reason for denying medical chimerism.

Might there be a secular argument against chimerism? In their thorough and sensitive article on the subject, Jason Scott Robert and Françoise Baylis appeal to the "yuck" factor. They perceive in our culture a prearticulate sense of repugnance, reminiscent of Leon Kass's concerns. Even though they acknowledge that the definition of a species is up for grabs and that "the unique identity of the human species cannot be established through genetic or genomic means,"[16] we confront an intuitive objection to chimerism. A fear is at work, a vague yet palpable fear. What is this fear? The fear of moral confusion. "The issue at the heart of the matter is the threat of inexorable moral confusion."[17]

Has the ancient Greek myth of the chimera returned in an oblique form? Do we have an imaginary fear to contend with? Are we terrorized by moral confusion?

Monkey Minds and Human Dignity

This brings us to the focus of the chimera debate, namely, mixing genes between humans and NHPs. We note that grafting stem cells into a human person's brain for experimental purposes is a "no-no." Behind this no-no, of course, lies the doctrine of human dignity. Accordingly, we treat each human person as an end and not merely as a means. Therefore, such human experiments would reduce a person to a mere means, and this is unethical. Scientists turn then to nonhuman models, because the dignity problem does not arise. Or, does it?

It might help us to understand what's going on here if we go back to the German philosopher of the Enlightenment, Immanuel Kant. It was Kant who provided us with the ethic of dignity that virtually the entire Western world now takes as axiomatic: We must treat a person as an end and not merely as a means toward some further end. But, this is not what is relevant here. What is relevant is that when Kant tried to figure out what a person is, the concept of rationality was introduced. A person is an animal with the capacity to reason. There is something special about reason, very special. "The ground of this principle is: *Rational nature exists as an end in itself*."[18] Theologians before and after Kant have been quite sympathetic to this, even to the point of connecting the immortal soul with the rational mind. Although the authors of the present book are quite skeptical about this equivalence, it is the prevailing assumption.

Contemporary Kantians have broadened the meaning of mind, so to speak, in their search for what is distinctively the human advance along the evolutionary trail. An opposable thumb? A bigger brain? A more complex brain? Language? Rich relationships? Morality? Culture? Spirit? Because higher primates exhibit some of these attributes, we think of them as close to human beings. At what point, we ask, do they cross a line and join us with the moral status of persons with dignity?

Because of our contemporary association of the brain with the mind, questions regarding brain cells will pop up in the context of such moral deliberation. Hence the focal question of the *Science* ethics team of Ruth Faden et al.: What might the implantation of human brain cells in a primate do to the animal's brain, and could this change the animal's moral status? If we place neuronal stem cells into the brain of a chimp or gorilla, might the animal develop a rational or immortal soul? "One conceivable result of H-NHP neural grafting is that the resulting creature will develop humanlike cognitive capacities relevant to moral status." And, if so, will we need then to afford to the animal all the rights that persons can claim in our society?[19]

It is important to recognize that the evidence to date indicates that the risk of this happening is low, if not nonexistent. Previous introduction of neural progenitor cells into developing animal brains led to integration of the human cells; yet, no animal behavior changes were reported. It appears that when hES cells are placed in primate brains, the host cells take over the direction of development. "Thus," concludes Cynthia Cohen, "evidence suggests that it seems highly unlikely that the transplantation of human neural stem cells into mouse or primate embryos would result in the development of nonhuman hosts with a human brain."[20]

In addition to evidence thus far regarding the noneffect upon animal brain function, the criteria for identifying distinctively human behavior produced by brain activity is not clear. So, assessment might be difficult. Overall, the *Science* ethics team concludes that it is "unlikely that the grafting of human cells into healthy adult NHPs will result in significant changes in morally relevant mental capacities." Yet, having said this, the team proceeds to express dramatic caution regarding the great apes. Why? Because the great apes, of all the animals, are already so close to humans in their genetic and behavioral makeup. Perhaps the risk is higher, some members of the ethics team surmised, that introduction of human neuronal cells could bridge the short distance from animal behavior to human behavior. The group's conclusion is that it supports the recommendation by the National Academies of Sciences that "H-NHP neural grafting experiments be subject to special review." In sum, because we lack the knowledge regarding what might happen, we will invoke a policy of precaution.

Such precaution could become an appropriate ethical guide for research practices. Scientists who plan to insert human embryonic stem (hES) cells into animal models could take a step-by-step approach. Using only the minimal quantity of human cells necessary for the experiment, the results could be monitored. Then the number could be increased incrementally until conclusions could be drawn.[21] If at some stage the fear is confirmed and an animal model begins to exhibit human capacities, then further steps could be abandoned. Precaution could become the laboratory guide.

Should HNH Chimeras Breed?

As mentioned above, following NAS guidelines researchers try to prevent affecting the germ line leading to the breeding of chimeric children. Why? What might happen if a human/nonhuman race is born? It would elicit the yuck factor. It would seem to violate human dignity. It would be repugnant to Leon Kass, and probably to the rest of us as well.

Here Leon Kass may have the better part of wisdom. We view babies as coming into the world as a blessing and a treasure. Human birth should take place within a loving family, where each child can be nourished and cared for. Procreation is a communal event with communal moral parameters. Human birth is much more than merely a physical event. To produce chimeric children as a by-product of scientific research would lead to individuals in our world with no home. The nest within which each newborn receives physical, emotional, and social care would be absent. "To develop nonhumans that might harbor human sperm or eggs and, in theory, could mate to create human children, would violate our sense that our children should not be produced as unanticipated side-effects of scientific studies," concludes Cynthia Cohen.[22] We concur.

Conclusion

Laboratory researchers and attending surgeons simply must employ human-to-human, human-to-animal, and animal-to-human transfers of genetic material to proceed with their daily work. To call this "chimerism" and then forbid it categorically would be tantamount to cutting off the electricity and water supply to the laboratory. If the imaginary fear of the chimera spitting a fire of moral confusion leads to the unnecessary prohibition of potentially valuable research and therapy, then the myth will have done more damage than the chimera could have done.

We believe human dignity is well worth protecting, to be sure. Yet, a threat to human dignity is not likely to arise from placing stem cells into animal tissue, or from placing animal tissue into human beings. Precaution, not prohibition, is the road reason ought to take through the wilderness of moral confusion.

Still, a matter we need to contend with is the yuck factor that rises up when thinking about chimerism. Yuck arises over the issue of patenting as well. In the next chapter we take up the issue of gene patenting—actually, the patenting of hES cell lines—known in the field as "intellectual property." Moralists working within the human protection framework frequently find this topic disgusting, because the very idea that someone could patent something so natural as the genome and related DNA phenomena elicits a feeling of repugnance. Yet, more is at stake than merely the relationship of life to intellectual property. At stake also is the need for big science to draw investment capital and for the ethicists to pursue economic justice.

Notes

1. National Academies of Sciences, *Guidelines for Human Embryonic Stem Cell Research* (Washington, DC: National Academies Press, 2005).

2. The National Council of the Churches of Christ in the USA, "Fearfully and Wonderfully Made: A Policy on Human Biotechnologies," www.ncccusa.org/pdfs/BioTechPolicy.pdf, lines 370–71.

3. Portions of this chapter draw upon earlier publications: Ted Peters, "The Return of the Chimera," *Theology and Science* 4.3 (November 2006): 247–60; and Ted Peters, "A Theological Argument for Chimeras," in *Nature Reports Stem Cells online at* www.nature.com/stemcells/2007/0706/070614/full/stemcells.2007.31.html.

4. Bible readers might recognize elements in this story. This is what happened to Joseph in Egypt. In that case, Potiphar's wife tried to seduce Joseph, the migrant from Israel. When Joseph refused, she told the same reverse lie to her powerful husband, who was an assistant to the pharaoh. In the biblical version (Genesis 39–41), the ruler placed Joseph in jail for a term. Then Potiphar promptly died, leaving Joseph without a court record regarding his release. Not until a new pharaoh ascended the throne did rescue from prison become a possibility for the Hebrew Joseph.

5. This plot is reminiscent of another biblical story, a story that also includes seduction. When the king of Israel, David, wanted to seduce the beautiful Bathsheba, he needed to remove an obstacle, namely, Bathsheba's husband, Uriah. As king, he was executive over the draft. So, he ordered that Uriah be drafted into the army and given a dangerous assignment on the military front, knowing the new recruit would see fierce battle. The plot worked. Uriah died in battle. The Jerusalem king took the new widow into his own harem (2 Samuel 11–12).

6. Hans-Georg Gadamer, *Wahrheit und Methode* (Tübingen: J.C.B. Mohr, Paul Siebeck, 1965), English trans. by Joel Weinsheimer and Donald G. Marshall, in *Truth and Method* (New York: Continuum, 1994), 301.

7. Jason Scott Robert and Francoise Baylis, "Crossing Species Boundaries," *American Journal of Bioethics* 3 (Summer 2003): 1–13.

8. A chimera is a creature with DNA, cells, tissues, or organs from two or more individuals. If the tissue comes from two different species, this produces an interspecific chimera. Chimeras are not produced through sexual reproduction, as hybrids are. Mules, born from a male donkey and a female horse, are hybrids, not chimeras. Helpful definitions are found in the document entitled "Hybrids and Chimeras: A Consultation on the Ethical and Social Implications of Creating Human/animal Embryos in Research," published by the Human Fertilisation and Embryology Authority of Great Britain, April 2007 (www.hfea.gov.uk).

9. Henry T. Greely, "Defining Chimeras and Chimeric Concerns," *American Journal of Bioethics* 32 (Summer 2003): 17–20.

10. N. Uchida, et al., "Direct Isolation of Human Central Nervous System Stem Cells," *Proceedings of the National Academy of Sciences USA* 97 (2000): 14720–25.

11. Mark Greene, et al., "Moral Issues of Human-Nonhuman Primate Neural Grafting," *Science* 309 (2005): 385–86.

12. Pontifical Academy for Life, *Prospects for Xenotransplantation: Scientific Aspects and Ethical Considerations* (September 26, 2001), n. 10; reprinted in *National Catholic Bioethics Quarterly* 2.3 (Autumn 2002): 481–505. Perhaps one might cite the ancient Hebrew proscription against bestiality

13. Duane T. Gish, *Evolution: The Fossils Still Say No!* (El Cajon, CA: Institute for Creation Research, 1995), 34. When creationists weigh in on the stem cell debate, they side with the embryo protectionists against hES cell research. See sss.icr.org/index .php?module=news&action=view&ID=18.

14. Gish, *Evolution*, 35.

15. Henry M. Morris, *Scientific Creationism* (Green Forest, AZ: Master Books, 1974, 1985), 209.

16. Robert and Baylis, "Crossing," 4.

17. Robert and Baylis, "Crossing," 10.

18. Immanuel Kant, *Groundwork of the Metaphysic of Morals*, tr. H. J. Paton (New York: Harper, 1948), 96.

19. See, for example, Robert Streiffer, "At the Edge of Humanity: Human Stem Cells, Chimeras, and Moral Status," *Kennedy Institute of Ethics Journal* 15.4 (2005): 347–370; also Phillip Karpowicz, Cynthia B. Cohen, and Derek van der Kooy, "Developing Human-Nonhuman Chimeras in Human Stem Cell Research: Ethical Issues and Boundaries," *Kennedy Institute of Ethics Journal* 15.2 (2005): 107–134.

20. Cynthia B. Cohen, *Renewing the Stuff of Life: Stem Cells, Ethics, and Public Policy* (Oxford: Oxford University Press, 2007), 131.

21. Cohen, *Renewing*, 132.

22. Cohen, *Renewing*, 138.

CHAPTER SIXTEEN

~

Justice and the
Patenting Controversy

It was the last day of an international conference on the Human Genome Project, held in Nice, France. Indeed, it was the last session of the last day. By this time, normally seats would be vacated, as conference-goers begin to drift toward home. Energy would be low, as three days of lectures and discussion drains attention from even the most stalwart of conference attendees. But on this day, even at the very last, energy was high. Indeed, as the speaker began, one could feel the energy level in the room rising. With it rose tempers. Why?—because this speaker was talking about efforts to patent portions of the human genome. The tension in the room became palpable. The United States, which was spearheading the patenting effort, came under severe criticism from representatives of other countries.

The energy that arose around patenting in the context of efforts to map the human genome continues today in the context of stem cell research. The question of intellectual property protection is complicated, and we cannot resolve it in this small volume. But it seems appropriate to offer at least a few reflections on how patents, and public opinion and energy around them, are affecting stem cell research. The issue is particularly important for us as Christian thinkers because in 1995, before the announcement of the isolation and development of human embryonic stem (hES) cells, representatives of some 80 religious organizations signed a Joint Appeal against Human and Animal Patenting. The Joint Appeal declared flatly: "We believe that humans and animals are creations of God, not [of] humans, and as such should not be patented as human inventions."[1] The signing of this petition may lead some people to believe that the only position compatible with religious be-lief is a position that denounces patents on stem cells. However, as we shall argue, there are both good reasons to permit patents and also some reasons to be cautious about their scope and application.

Patenting Stem Cells

Prior to 1980, U.S. courts considered living matter to be "preexisting" in nature and therefore not something that could be patented. However, in *Diamond v. Chakrabarty*, things changed. Chakrabarty had created a new form of bacteria. The Supreme Court found that it was therefore a "human invention" even though it was living matter. They permitted the patent and thereby laid groundwork for a new approach. Since then, corporations such as Human Genome Sciences of Rockville, Maryland, have applied for and received dozens of patents on DNA and human-derived tissues.

Based on this established line of reasoning, when James Thomson isolated and cultured human embryonic stem cells in the laboratory, the Wisconsin Alumni Research Foundation (WARF) applied for and received several key patents on Thomson's work. In its application, it sought a patent on "A purified preparation of primate embryonic stem cells which (i) is capable of proliferation in an in vitro culture for over one year; (ii) maintains a karyo-type in which all the chromosomes characteristics of the primate species are present and not noticeably altered through prolonged culture; (iii) maintains the potential to differentiate into derivatives of endoderm, mesoderm, and ectoderm tissues throughout the culture; and (iv) will not differentiate when cultured on a fibroblast feeder layer."[2] The patent office examined these claims and granted several key patents to WARF.

As is typical when patents are granted, these patents give WARF the right to exclude everyone else in the United States from making, using, sell-ing, offering for sale, or importing any hES cells covered by these patents to the year 2015. These patents not only granted protection to WARF for the *method* of deriving stem cells, but also give WARF protection for the *composi-tion of matter*—that is, for the stem cells themselves. Regardless of how one obtains hES cells, those cells are covered by the patents. The result is that any use of this composition of matter—hES cells—is subject to the license-granting authority of the WARF.

Those who felt they would be suffocated by this arrangement have pro-tested, leading to a 2007 decision by the U.S. Patent and Trademark Office (PTO) to revoke three WARF patents. The PTO's action was a reaction to protests filed by Jeanne Loring, a biological researcher at Burnham Institute for Medical Research in La Jolla, California, along with the Foundation for Taxpayer and Consumer Rights in California, as well as the Public Patent Foundation in New York.[3] As of this writing, WARF has a right to an appeal; the patents remain in effect pending this appeal.

Are Patents Justified?

Are there some things that should *not* be patented? Have patents been stretched too far when they include embryonic stem cells? The patenting of stretches of human DNA has been extremely controversial. The patenting of embryonic stem cells is also controversial, as the recent challenges demonstrate. While it is not our purpose to explore this debate fully, we do offer some reflections here on the arguments for and against patenting of human body parts, tissues, and tissue derivatives such as embryonic stem cells.[4]

The basic arguments *for* allowing patents in general are two. First is a utilitarian argument: patents are necessary for scientific progress. For instance, William Haseltine of Human Genome Sciences, declares: "The patent system is . . . structured and administered to ensure the rapid and open dissemination of new knowledge, encourage innovation, and promote commerce." The idea is simple: without a patent, companies would be afraid to share breakthroughs such as the derivation and isolation of stem cells; with patent protection, they share knowledge and ultimately everyone is better off. It is a classic utilitarian argument: patents are justified because they do the greatest good.

But there is a second argument as well, based not on long-term utility but on a presumption of ownership "rights" over the fruits of one's labors. For instance, the International Covenant on Economic, Social, and Cultural Rights (in force since 1976 and signed by more than 130 nations) provides that every person has a right "to benefit from the protection of the moral and material interests resulting from any scientific . . . production of which he is the author." An inventor has a right to some of the proceeds and profits from his or her invention. This notion derives from John Locke's view that people should have a right to the proceeds of their labor. Such a view may of course be compatible with utilitarian arguments, on grounds that the greatest good is done overall by permitting people to have rights to their inventions. However, the argument is not strictly speaking utilitarian, for one might have a right to profits from one's invention even *if* permitting those profits does not foster the general good or is not the fastest way to overall benefit. This distinction in arguments becomes important in the stem cell arena, for it leads to disputes about whether patents on stem cells are permissible only if they serve the general good, or whether they are permissible as a reflection of the rights of inventors.

The arguments *against* patents generally focus on the utilitarian concerns. They include the charge that patents do not *support* scientific research and advancement of knowledge, but rather put roadblocks in the way. This is certainly what many scientists think. It appears to be the primary reasoning behind the challenges to WARF's patents on hES cells.

However, as we saw above, opponents of patents make other arguments as well. They charge that human bodies and other forms of living matter are somehow "sacred" and therefore should not be patentable, nor should their derivatives. This would seem to be the underlying argument of the 1995 Joint Appeal against Human and Animal Patenting, cited above. By arguing that "humans and animals are creations of God, not [of] humans," the signatories appear to suggest that human and animal life has a "sacred" quality that should hold it inviolable from certain kinds of human intervention or claims, such as patenting.[5] Of course, this, too, may simply be a utilitarian argument: to permit patents on human tissues is ultimately not the "greatest good"; rather, the greatest good is protecting the sanctity of human tissues. However, the argument is not necessarily utilitarian; it seems rather to be a claim that some things can *never* be done, no matter how much "good" would ensue. It attempts to hold "sacred" and exempt from cost-benefit analysis certain kinds of things, including human tissues and their derivatives. Even nonreligious people often balk at the idea that anything that seems like a human body part should be subject to being patented.

What is often neglected in such arguments against patents is the question of "rights" to the results of one's labor. This mismatch may be why the debate has become so stalemated. Because the arguments *for* patents are not simply utilitarian, but are also based in considerations of rights, arguments *against* patents that look only at the question of good and evil seem to miss the mark.

Some Complications

Any effort to take account of rights, however, is complicated today. The Lockean notion of rights to the products of one's labor deals with an image of a single inventor claiming rights to his or her property. Today, the situation is far more complicated. Scientists today do not usually work in isolation. They are generally employed either by corporations or by universities. Under contract with their employers, they rarely "own" the results of their labor; rather, those results are considered products of the corporation or of the university. Moreover, research at universities is supported often by grants from the National Institutes of Health (NIH), but increasingly by grants from private corporations. Quietly over the last few decades commercial interests have step-by-step encroached on—some would say, conquered—university life. The quantum jumps in corporate funding for university research have multiplied, and so have the strings attached. Universities have instituted offices for patenting and licensing operations to market inventions by their faculties. Some critics charge that we have created a giant "academic-industry complex" in which professors look increasingly like business tycoons, and universities look increasingly like corporations investing in patents and profit.

This is the view taken by Jennifer Washburn in *University Inc.: The Corporate Corruption of Higher Education*. The word "corruption" in the title alerts the reader to Washburn's conclusion. Washburn argues that a tension between the values of academia and of business has been healthy for our society. Academia represents pursuit of knowledge, service to humanity, and cultivation of well-rounded persons and citizens. Business represents progress, production, and profit. The combination of the two value systems has contributed to the strength of our robust society. To eliminate one or the other would count as a great loss. "Although the profit motive plays an important role in our society, so do other values that limit and constrain what unregulated markets will do if left to their own devices."[6]

For Washburn, then, "the single greatest threat to the future of American higher education [is] the intrusion of a market ideology into the heart of academic life."[7] Universities and their associated private foundations have benefited from the 1980 federal Bayh-Dole act, which permits universities funded by federal funds to file for patents and keep the profits. Of course, the original intention was to maximize unlicensed and inexpensive usage of new discoveries. The original aim was to protect the open and free flow of research information. The intrusion of business interests, with its profit motives, threatens this free flow, in Washburn's view. She believes that we have poisoned the well where we used to go to quench our thirst for knowledge. The previous campus attitude of openness and sharing has been replaced by a proprietary subculture of industrial secrecy. Further, she sees it as a fundamental injustice: "The tragedy is all the more egregious when one recognizes that taxpayers are footing the bill for so much of this upstream research—paying, in effect, for the building and maintenance of a public library of knowledge to which they themselves are subsequently denied access."[8]

One need not go as far as Washburn does to see that there are complications and difficult ethical issues involved in current arrangements. James Thomson was employed by the University of Wisconsin. WARF is the private corporation that handles patents related to research conducted at the university. Thomson's research on hES cells was funded by the Geron Corporation of Menlo Park, California, which also funded John Gearhart's research on human embryonic germ cells and research conducted at the University of California, San Francisco, by Roger Pedersen. The funding of this research was part of Geron's commitment to study causes of aging and ways of alleviating the diseases of aging. Its primary research was on telomerase, and its support for human stem cell research was an outgrowth of prior discoveries of the importance of telomerase expression in mouse embryonic stem cells.

Thomson's hES cell work therefore touched on not only his own rights as an inventor, but on the rights of the institution that employed him and the corporation that funded the specific research. This is why WARF rather than Thomson himself applied for key patents on Thompson's discovery. This is also why the Geron Corporation has a special relationship to the patents now held by WARF. In a carefully arranged agreement, WARF emerged with the patent and Geron emerged with an exclusive license to research three cell types that could be derived from hES cells: neural cells, cardiomyocytes, and pancreatic islets. Any others who wish to work in these specific areas must negotiate with Geron to set "reach through" fees and royalties. All of Geron's other rights, however, are nonexclusive.

As this small example shows, the question of patent ownership is far more complicated than the "one author, one invention" model that undergirded original patent law. Who should hold a patent? Should it be the scientist? The hiring organization? The sponsoring corporation?

In the case of hES cells, the situation is further complicated by the federal restrictions on embryonic research, reported in an earlier chapter. As we noted there, President George W. Bush's announcement on August 9, 2001, permitted federal funding for research on existing stem cell lines but not for the derivation of new lines. The president clearly thought that at least 60 characterized or usable lines existed. However, researchers knew that worldwide only a dozen or perhaps up to 20 had been characterized. All of these were derivatives of the Wisconsin lines. Anyone wanting U.S. government funding for stem cell research would therefore have to negotiate with WARF. Washburn reports that one witness in a congressional hearing chaired by Senator Arlen Specter, R-Penn., feared that WARF's monopoly on hES lines would mean that nearly half of all U.S. government money granted in this area would be diverted to the University of Wisconsin and Geron in the form of royalty payments.[9]

In order to facilitate research, the U.S. NIH and WARF worked out a "memorandum of understanding" (MOU), according to which WARF would retain the rights granted by the 1998 patent whenever hES cell lines would be used by NIH-funded researchers. In return, WARF agreed not to impose restrictive costs on not-for-profit researchers and gave free access to stem cell lines to laboratories employed by NIH, the Food and Drug Administration, and the Centers for Disease Control and Prevention in Atlanta.

In other words, the possession of the patent does not impede basic research, as cell lines are given without cost for this purpose. When downstream products leading to therapy are developed, however, then the WARF license royalty clauses kick in. Upstream access is coupled with downstream costs.

In early 2002 NIH made similar agreements with other labs that had derived lines, such as the University of California at San Francisco, MizMedi in Korea, BresaGen in Australia, Technion in Israel, Cellartis in Sweden, and ES Cell International in Singapore. NIH awarded these institutions grants of $200,000 to $500,000 per year to develop an infrastructure for extending research based upon the WARF agreement. Their respective MOUs limited future charges for use of a stem cell line to $5,000 per line for U.S. labs and $6,000 for foreign labs, the extra to cover shipping costs. WARF now requires a license agreement with any and every academic laboratory inside the United States, whether or not they appear on the NIH registry. As for private biotech companies, they may pay WARF as much as $125,000 upfront plus a $40,000 annual license maintenance fee.

All of these arrangements suggest that stem cell therapies will be big business. We believe that it is partly for this reason that objections have been raised to the patenting of hES cells. Although the specific *charges* opposing the patents may be that the method used for derivation was not sufficiently "new" to qualify for a patent,[10] the underlying *issue* may be money.

Conclusion

So where do we stand on the question of patenting hES cells? First, we would note that we do not consider hES cells to be "sacred" in and of themselves. The reasons for this are outlined clearly in chapters above. Thus, we do not find compelling the arguments that suggest that stem cells should not be patented because they are forms of life that are sacred.

Second, we do believe that the general system of patents can serve the common good. Detractors often neglect the enormous start-up costs for innovation in biotechnology. In the case of hES cells, for instance, it will take years and years of basic research before any "therapies" are developed. What provides the impetus for all that expenditure into research is often the hope of financial gain at the end. Biotechnology companies must find investors willing to expend enormous "up-front" costs in hopes of eventual benefits. It is in the interests of all of us for scientific research to proceed. To the extent that the patent system helps to protect and encourage research, it is justified.

However, we also argue that because patents are justified by reference to the common good, they can be overridden by concerns deriving from the common good. For this reason, it seems important to us that intellectual property protection is carefully circumscribed and not granted too broadly, especially on upstream discoveries.

Third, we honor rights of ownership deriving from "invention." For instance, we expect to receive royalties on this book, which is our "invention." However, we also believe that inventors' rights are not absolute; they, too, can be overridden by concerns for the common good.

Those concerns need to balance profits and basic human needs, return on investment and basic justice. The desire for return on investment is natural. The desire for profit is also natural. However, as 1 Timothy 6:10 (NRS) says, ". . . the love of money is a root of all kinds of evil." We note in this Bible passage that it is the "love" of money that leads to evil, not the money itself. We further note that our human life is ambiguous. Money as money can do good things, especially when pressed into the service of higher goods. We believe human health and well-being constitute higher goods. Despite the drawbacks of the "love of money," we believe the expenditure of money to support stem cell research provides an opportunity for a leap forward in medical therapy that would not be possible without it. Christians have a word for this. We call it *stewardship*. Protections that ultimately serve the common good are a form of stewardship and can be supported. What is crucial, however, is a constant examination of whether the common good is indeed served by the patent process in general and by particular patents such as those granted to WARF in the stem cell arena.

Finally, the issue of patents can be folded into the larger concern over economic justice. In the stem cell controversy, concern for justice takes the form of concern for access, access to the fruits of the labor of medical researchers. Once downstream products offer therapies of regenerative medicine to the public, who will have access? It is the long-term financial impact of patents that becomes the moral issue. From within the future wholeness framework we ask: will the patenting of stem cell lines and their products prevent access by the poor around the world or contribute to it? As Christian ethicists, our goal is to see that as many hurting people as possible gain access to the hope that regenerative medicine brings.

Notes

1. C. B. Mitchell, "A Southern Baptist Looks at Patenting Life," in *Perspectives on Genetic Patenting: Religion, Science and Industry in Dialogue*, ed. A. R. Chapman (Washington, DC: American Association for the Advancement of Science, 1999), 169.

2. Cited by Jeanne F. Loring and Cathryn Campbell, "Intellectual Property and Human Embryonic Stem Cell Research," *Science* 311 (March 24, 2006): 1716–17.

3. See Erika Check, "Patenting the Obvious?" *Nature* 447 (May 3, 2007): 16–17.

4. For a fuller exploration, see Karen Lebacqz, "Who 'Owns' Cells and Tissues?" *Health Care Analysis* 9 (2001): 353–67.

5. For a full discussion of the 1995 controversy over the Joint Appeals against Human and Animal Patenting, see Ted Peters, *Playing God? Genetic Determinism and Human Freedom*, 2nd ed. (London: Routledge, 2003), chap. 5.

6. Jennifer Washburn, *University Inc.: The Corporate Corruption of Higher Education* (New York: Basic Books, 2005), xviii.

7. Jennifer Washburn, *University Inc.*, x.

8. Jennifer Washburn, *University Inc.*, 155.

9. Jennifer Washburn, *University Inc.*, 151.

10. Cathy Tran, "WARF Stem Cell Patents Challenged" (www.the-scientist.com/news/display/25037.

CHAPTER SEVENTEEN

~

The Spiritual Soul
and Human Dignity

The various positions taken in the worldwide controversy over stem cell research—both overtly religious and covertly religious within the secular—begin with assumptions regarding the human soul. Commitments to human dignity are based on assumptions regarding the nature of the soul. Perhaps it is time for us to turn to the theology of the soul to see if we can shed further light on what is at stake.

We will share our conclusion first, then turn to our analysis. We conclude that the term *soul* refers to our inmost essence as an individual self, while the term *spirit*, which overlaps with *soul* to be sure, refers to our capacity to relate with one another and with God. While the word *soul* connotes who each of us is as an individual, the word *spirit* connotes that dimension of our personal reality that unites us with others. We also believe that the terms *soul* and *spirit* should refer to our existence as embodied creatures. The soul is not a ghost-like entity that simply inhabits a body. Rather, to speak of soul reminds us that as embodied creatures we have a center of identity, a centered self. To speak of spirit reminds us that we exist in a web of real relationships. The power of the spirit unites us with something bigger than us. *Souls as centered selves are formed by and develop in spiritual relationships!* Our soul only becomes immortal because of our spiritual relationship with the eternal God. To think of a human person in the fullest sense is to include body, soul, and spirit in relationship to community and to God.

Now, to our analysis. How did we reach these conclusions? It is our considered judgment that too often the public debate over stem cells operates with

207

only one theology of the soul, only one selection on a menu of theological options. Embryo protectionists, in their effort to protect what they perceive to be innocent lives under attack by dangerous Frankensteinian scientists, appeal to the doctrine of the soul to justify their protection of human dignity. Although we agree that the Christian doctrine of the soul is crucial to understanding the stakes of the stem cell debate, we disagree with the assumption often at work in the embryo protection position, namely, that the cells of the blastocyst in the laboratory are ensouled and therefore sacred. We disagree with the assumption that if one is for the soul one must be against embryonic stem cell research. Such assumptions cover over alternative understandings of the soul—understandings that are equally if not more coherent and more credible. In our judgment, the concept of the soul with which the Vatican and some evangelicals work is insufficient. Our task in this chapter is to explain why while providing an alternative vision of the human person.

We believe theologians can shine new light on the controversy and elicit renewed compassion for suffering persons by providing an understanding of the human soul that is often omitted from the stem cell debate. We argue that the concept of the soul—and therefore the concept of human dignity as well—depends on our relationship to God. God graces the human person with divine love, and this grace testifies to the person's ultimate worth. Rather than think of the soul as a substance or thing to be protected in the developing blastocyst we think of our soul and spirit together as the centered opening within us that allows us to enjoy God's everlasting presence.

Is the Soul a Spiritual Substance?

The first question we need to pose is this: Are those who defend the early embryo from destruction by laboratory scientists assuming substance dualism? Is the view put forth by the Vatican and other embryo protectionists dependent upon a specific metaphysics that deems the soul to be a spiritual substance added to our material bodies? Does human dignity depend upon such a spiritual substance? Does the moral protectability of the blastocyst depend upon the presence of a spiritual substance we know as soul? Answers to these questions might help us to understand just what is at stake and just where, if anywhere, we might find sufficient wiggle room to pursue reconciliation of differences between contenders in the stem cell debate.

The first thing a substance dualist believes is that the soul is a spiritual substance. The sould is an immaterial and spiritial thing, described as *forma substantials*, to use the Vatican's voculary.[1] Furthermore, we human beings are said to be a composite of two substances, one physical and the other spiritual. Even though the spiritual soul resides *with* the body, it is not itself bodily. Like pouring a few drops of motor oil into a bowl of water and watching the colors swirl around, the two substances are located in the same place but they do not mix. Still further, in the dualist view, who we are in essence is found in our spiritual soul, not our physical body. In fact, when we die, our essence will continue to live on while the physical body is entombed in the cemetery. The spiritual soul is the immortal soul.

It is in the soul where our mental activities take place, where our identity lodges. Although plants and animals have souls in a primitive sense of a life force, the highest form of soul, according to the ancient Greeks, is the rational mind. The Greek word from which we get our words psychology and psychiatry, ψυχη or *psyche*, can be translated as either "soul" or "mind." A chief spiritual goal in the ancient Greek-speaking world was to overcome the struggle between our higher spiritual or mental nature and our lower physical or bodily nature. The appetites of the body can distract and disorder the mind, to make it *zerstreut* (a German word meaning disintegrated, scattered, absent-minded). So the spiritual task was one of self-discipline in which the mind would master the body's passions. This is reflected in the background to the New Testament where St. Paul speaks of the battle between flesh and spirit. From this spiritual struggle we get modern phrases such as "mind over matter."

The idea of an individual soul seems to fit with our everyday experience.[2] We emphasize *seems* to fit our experience. What we experience is that we operate with a single centered consciousness. As we think about what we see and hear and smell, there is a single "I" or "me" that is the focus of this thinking. Even when we have a hard time making up our mind and say that "I'm of two minds about this," as soon as we make a decision we become one mind again.

When we can conceive in our rational mind that to lose weight will improve our health, we may experience a struggle when walking the aisles of the supermarket trying to strengthen our wills to combat the temptation to buy and devour all that appears tasty. It seems reasonable to explain this experience as a battle between flesh and spirit, between our physical bodies and our spiritual souls.

The Hindus of ancient India formulated such a doctrine of the soul as a spiritual substance. The essential self or *atman* is buried within the body, buried in the darkness of physical strivings, passions, and actions. Bodies and bodily activities are plural, divided, multiple; whereas the soul is unified, one, singular. The body is dark, and the buried soul within us seeks to unbury itself, to rise up toward the light and unite itself with universal truth.

Enlightenment—that is en*light*enment—is the spiritual goal of Hindus. Once we have purified our soul from bodily contamination and attained enlightenment we will realize that our soul is not finally an individual substance. No, it is at one with the whole of reality, the All; it is at one with *Brahman*.

In the meantime, for Hindus, while the soul is still contaminated by bodily actions (*karma*), the soul remains a single spiritual substance with its own independent destiny. It recycles. When we die, our soul proceeds to become reborn in someone else's body. This is the process of metempsychosis or the cycle of rebirth. We know it as reincarnation—that is, from *re* (again) and *carnal* (fleshly). Reincarnation means entering the flesh again and again. The final destiny of the soul for Hindus is to jump off the wheel of reincarnation, to leap into the realization of the oneness of *atman* and *Brahman*. "I am Brahman" (*Aham Brahmasmi*), says the Advaita sage, Shankara.[3]

What the wheel of rebirth signifies, according to Hinduism, is that the soul is eternal. Or, to be more specific, it comes from eternity and returns to eternity. While visiting our bodies in time, the soul becomes an individual substance. The soul is individual, to be sure; but it is individual only as long as it follows the path of incarnation and reincarnation. Each one of us has a soul, but our bodies pull it down into the darkness of physical attention and action to prevent it from liberating itself for a purely spiritual realization. When liberated from its attachment to physical substance, it returns to *Brahman* and escapes further rebirth.

When early Christians flirted with a version of substance dualism, they insisted on one item that sharply distinguishes their view from Hinduism and from similar versions of metempsychosis among the Greeks. The Christians insisted that the soul did not come from eternity. Rather, the soul began when our bodies began. The substance dualists of today who are Christian insist that God creates a brand new soul for each individual human body. The soul may last forever, but it has a beginning. And, the soul does not get recycled. One incarnation is enough.[4]

One early debate, a debate we can see taking place within the theology of St. Augustine of northern Africa during the end of the fourth century and the beginning of the fifth, is whether each soul is specially created or whether each baby inherits some aspect of his or her soul from the parents. This second view has been called *traducianism*. Augustine could not definitively resolve this in his own mind. The Bible is not clear on this, so Augustine could not be clear. Yet his followers for the most part have rejected traducianism and affirmed *creationism*. Creationism is the notion that God creates a unique soul for each person.[5] No official church dogma establishes creationism as the only Christian view; yet creationism was the voiced view of Pope John Paul II.

Is Pope Benedict XVI against Substance Dualism?

Was Pope John Paul II a substance dualist? Well, it certainly looks like it. And, it appears that his justification for protection of *ex vivo* blastocysts in laboratories depends upon just this view of the soul. If something has a soul, it is immortal. And, if it is immortal, it has dignity. And, if it has dignity, then we cannot exploit it for the ends of medical research. This seems the tight logic of the previous pontiff's argument.

Yet, just how dependent is the Vatican on substance dualism? Decades before being promoted to the large office with the view, Pope Benedict XVI, previously known as Joseph Ratzinger, said he was rejecting substance dualism. In those days Ratzinger was respected everywhere for his theological acuity, even his ecumenical charity. He studied and analyzed the concept of the soul. He concluded that the Christian understanding is unique, the product of its own intellectual development. What Christians believe is not merely a repeat of something said earlier by Plato or Aristotle, nor could it be reduced to the substance dualism found in any other philosophies.

Theologian Ratzinger emphasized that, even though the soul is distinct from the body, it contributes to the unity of the one person. It takes both a soul and a body in relational unity, centered in God, to make a person a person. We can speak of the human soul as immortal only because God has promised to raise us from the dead. God promises to raise us as persons complete with body and soul. Along with other contemporary theologians, Ratzinger has rejected the idea of an autonomous, substantial soul with a built-in immortality. "In none of the great theological teachers have I found a purely sustantialist argument for immortality."[6] In the place of substance dualism, he favors a perspective that regards God's decision and activity to raise human persons in the "body" as the distinctive Christian view.[7]

Now, we need to pose questions to the Holy Father. Just what is being rejected and what is being affirmed? Is he actually rejecting the idea that the soul is a spiritual substance separate from the body; or is he merely rejecting that this spiritual substance on its own is immortal? If he rejects substance dualism, is the ground for this rejection solely to defend "resurrection" as the means to immortality? What is less than clear is whether he also rejects the very notion of a nonbodily, immortal, spiritual substance. We still need to ask for greater clarity: What is the metaphysical status of the soul? Is it a substance distinct from the body? Or, is it the form of the body that becomes immortal only in the eschatological resurrection? If the latter, then it would cease to be the decisive factor determining moral protectability in a laboratory petri dish. It would fall short of making cells sacred.

If we are looking for wiggle room in the Vatican position, perhaps we can find some here. Pope Benedict XVI is not committed to substance dualism. If the strong stand within the embryo protection framework is dependent upon substance dualism in order to mark the time of ensoulment and hence the time of moral protectability, then perhaps an alternative concept of the soul might open some new doors to shared understanding.

A Theological Evaluation of Substance Dualism

We, the authors of this book, have difficulties affirming substance dualism. So, what's wrong with it? To answer we'll evaluate the idea of substance dualism according to the four criteria familiar to Methodists and Nazarenes. These four are known as Wesley's quadrilateral: Is it scriptural? How does it appear in Christian tradition? Does it fit with experience? Is it reasonable?

First, substance dualism is not scriptural. To be sure, the Bible talks of souls. Indeed, the Bible talks frequently about our soul; it talks about our spirit as well. It also talks about the Holy Spirit. But nowhere does it describe the soul as a *spiritual substance* especially created by God to make human reasoning possible. So our concern is not whether there is or is not a soul, but rather, whether substance dualism is the best interpretation of what the Bible says. We believe it is not.

A key passage is Genesis 2:7, in which God formed "a man (*Adam*—[an earth being]*) from the dust of ground (*Adamah*) and breathed into his nostrils the breath of life, and man became a living being (*nephesh*)." This Hebrew word, *nephesh*, referring to our animal life force, is translated as "soul" in English. Without soul, we cannot live. Who we are as living beings is a whole, a unity of body and soul. Hebrew scholars are reluctant to associate the Old Testament concept of *nephesh* with the immortal intellect of Plato.[8]

In the New Testament, according to St. Paul, the soul dies. It perishes (1 Corinthians 15:35–49). Our entire identity drops into nonbeing. St. Paul still believes in resurrection, to be sure. What is raised is our entire person, and the form of resurrection he calls a "spiritual body" (*soma pneumatikon*). Because of this, Christian creeds affirm "resurrection of the body," not immortality of the soul. In summary, the concept of the soul as a spiritual substance that is immortal is not what the scriptures teach.

Second, does substance dualism cohere with Christian tradition? The idea that the soul is a spiritual substance with immortality does appear in some places in later Christian tradition. Despite the fact that many contemporary theologians argue that the Christian tradition deviated from its biblical roots by adopting an alien Greek form of thinking about the soul, the fact is that the tradition includes substance dualism. At no point does it become dogma, yet it is true that over the centuries many Christians have gotten along quite well believing they possess immortal souls. So, we would be reluctant to go so far as to say that substance dualism and Christian belief are totally incompatible. Substance dualism is not heresy. Well, not quite. The question is this: What is the best understanding of the human soul? This is an urgent theological question, because just how we understand the soul has implications for the ethics of stem cell research.

Third, we as human persons do in fact have the experience of a unity of rational consciousness. We get anxious when on occasion we become flustered by overstimulation and cannot get all our thoughts together. To feel *zerstreut* is uncomfortable. Yet, barring mental illness, our rational minds seem to have the power to work through the threat of multiple stimuli and organize otherwise scattered thoughts until a unity of consciousness is regained. Our organizing center seems to have a way of exerting itself. Just how do we explain this? Could it be that our rational consciousness is due to the presence of a spiritual substance known as the soul?

This seems plausible to some. This leads one evangelical Christian dualist to write, "Scripture as a whole does not teach that the soul exists. Scripture simply presupposes the existence of the soul because its existence is affirmed by the common sense of ordinary people."[9] Note what this theologian is saying: Substance dualism fails the test of being biblical but passes the test of being experiential.

Yet, without denying the existence of the soul, we still ask: Might explanations other than dualism be equally good or even better at explaining our experience? One such alternative seems obvious. It is the observation that we have one body. We have one circulatory system, one nervous system, one set of limbs that work together, one heart, and one brain. Is it only an accident that one body would have one consciousness rather than two or more? The principle of parsimony—that is, the principle of selecting simpler explanations over more complex ones—would suggest that the singularity of the body is a better explanation for the singularity of consciousness than the addition of a second substance, a spiritual substance. This may not by itself be decisive for refuting substance dualism, but it counts as a counterargument.

Fourth it is reasonable to doubt substance dualism. However, despite what we just said, it is also reasonable to believe that our soul is a spiritual substance. Reason alone in the form of science or philosophy may not be able to decide. Don't be bullied by philosophers who smugly denounce religious ideas about the soul as prescientific.[10] What such complaints usually reveal is that the materialist philosopher—not the laboratory scientist—is a reductionist. The reductionist wants to reduce everything spiritual and everything personal to physical causes, such as genetic information or brain action. Although they claim science is on their side, a closer look will show that such reductionism is an ideology and not itself science. No disproof for substance dualism has been yet established using science alone.

Now, for just a moment, it may appear that we are defending substance dualism. Instead, we are simply trying to locate where the criticism should derive. Our criticism comes from theology, not philosophy and not science.

We see significant *theological* problems with substance dualism. And these theological problems have ethical implications that affect the way we approach the stem cell controversy. We have just dealt with four evaluative criteria: Scripture, Christian tradition, experience, and reason. Now, in a more summary fashion, we turn to our theological objections to substance dualism.

The Soul and the Mind

Let us continue beyond the four criteria of the Wesleyan quadrilateral to additional measurements of the position we dub substance dualism. Our fifth concern is this: We see a problem with the close identification of the soul with the human capacity to reason.[11] This equation of the soul with the rational mind comes from the ancient Greek philosophers, not from the Bible. Even though it partially squares with experience, to be sure, it fits only the experience of mature and healthy adults. It does not fit the experience of senior citizens whose brains have been damaged by Alzheimer's and who require constant care because their rational powers have been impaired. In such cases must we say they've lost their souls? Or, what about the baby early in the womb prior to brain development? No rational capacity exists here at all. So, does this imply the absence of a soul? Or, returning to the adult who experiences a unity of rational consciousness, we know that we can have a bad day. We can become so overcome with physical suffering or mental anguish that our reasoning capacity becomes compromised. Does this mean a loss of some of our soul's substance? When we lose our sense of center, does the spiritual substance vacate us and leave us with only a body, with only a physical substance? If so, will this remove our moral protectability? The coherence of substance dualism collapses under *reductio ad absurdum*.

If we identify the soul with the capacity to reason, then it means only some people have souls while others don't. It also means that we possess a soul for a period of time in our life when we are able to reason, while leaving inadequately explained just how we relate to our souls before we get a brain and when our reasoning capacity is impaired. If embryo protectionist theologians respond to this problem, they have available the notion of spiritual substance. They could say that the soul's substance is present in the embryo prior to brain development and in the postrational Alzheimer's patient; but to make this claim they would need to sacrifice the rational quality of this substance. So we can see that, in order to make the argument work, embryo protectionists need to draft auxiliary supports to prop up the position. Eventually substance dualism will simply collapse under the weight of its own need for too many qualifications.

Having said this, we perceive an even more grave difficulty with substance dualism. It is the difficulty of identifying the soul with any innate human capacity whatsoever. Whether identifying the soul with reasoning or any other human ability, such an anthropology risks obscuring another extremely important Christian principle, namely, God relates to us graciously. God loves us before we can love ourselves, let alone love God in return. God approaches us, calling us into being and into relationship with God. God's call to us is not dependent on some capacity we have, nor even on a capacity God would give us along the way. To anchor our relationship with God in one aspect of human being, in the spiritual soul, risks establishing the value of each person on the value of the innate soul. In contrast, we believe the Bible teaches us that our value is due totally to God's valuing of us. "For God so loved the world," we see in John 3:16, "that he gave his only begotten son that whoever believes in him will not perish but rather enjoy eternal life." The alleged presence of a spiritual substance adds nothing to this initiative on God's part, nor is it necessary to understand eternal life with God.

It is our view that when we use the word "soul" we should elicit thoughts about our center as a person, and we believe we should also include the dimension of spirit so we can see that we have a relationship to a gracious God. Our soul is our essence, and our centered essence is present even when we fail to experience it rationally. It is present because God is responsible for it. Our essence is determined by our relationship to the God who calls us into being and with whom we will have an everlasting sharing. The human soul is not an immortal substance. Rather, what we might call "immortality" or life beyond death will be due to a creative act on the part of the God who loves us. God will act to raise us from the dead to live eternally within the divine life. Our future resurrection is in no way related to possessing now a spiritual substance that is immune to death. Rather, our resurrection is to-tally dependent on a future act of God, the act whereby God will fulfill the Bible's promise to raise us into the new creation.

The Soul in Stem Cell Research

When our theological position becomes applied to the ethics surrounding the blastocyst, we do not look for a moment of metaphysical ensoulment. We do not look at conception as a moment when God creates a new spiritual substance and attaches it to a newly established genome. Rather, we look for a relation-ship that calls, cultivates, and nourishes a nascent life toward individuality and communality with both the mother and with God. The random fertilization and flushing of ova within a woman's body does not qualify as such a relation-ship; nor does the appearance in a petri dish of a zygote produced either by *in vitro* fertilization (IVF) or sematic cell nuclear transfer (SCNT) qualify. What counts is the day a woman realizes that living within her body is the seed of a new life, a new life that she (and her partner) will welcome into this world, a nascent person whom the angels will ferry into everlasting life with God.

We start theologically with the eschatological future, with the divinely promised destiny for each citizen of God's kingdom. Then we look backward. We ask, when does this journey from nonbeing into temporal being and into eternal being begin? It cannot begin before there is a relationship that is life-giving, nourishing, and cherishing.

We distinguish between a potential person's relationship to his or her par-ents and the relationship with God. Parents who lack dispositions of loyalty and love toward their child do not in any way compromise God's love for that child or modify God's plan for salvation. Yet, working backward from our vision of salvation, we see the story beginning when the relationship begins that makes possible the development of centered personhood and enjoyment of communion with God.

With these theological considerations in mind, we find ourselves closest to those ethicists who support the 14-day Rule or, better, associate the onset of individuality and inviolable dignity with adherence to the mother's uter-ine wall and the appearance of the primitive streak. When this relationship between a mother and a future individual child is established, to think of the possibility of personhood and the prospect of a future destiny with God makes sense.

We cannot claim our position is stated explicitly in the Bible; nor can we claim that it has been long supported by Christian tradition; nor can we testify that it explains anybody's experience. But, our position is a reasonable interpretation of the Bible's message of salvation applied to this question: When can we speak of a human soul in such a way that it applies to the life of the early embryo?

Again, our point here is not to establish when an elective abortion—abortion understood as the removal of a fetus from a mother's womb—is licit or illicit. Rather, we offer the more modest point that substance dualism ought not to be relied upon as a theological foundation for an ethic that forbids employment of *ex vivo* blastocysts for stem cell research.

The Future of Dignity

The lynchpin of the embryo protection argument articulated most clearly by the Vatican is the insertion of dignity at the moment of conception. Once dignity is imputed to the zygote, then it is morally protected from destruction at the hands of laboratory technicians. The soul makes the zygote sacred. This sacredness gives the early embryo dignity. With dignity comes a "No Trespassing" sign to laboratory scientists.

This Vatican argument depends upon ensoulment, and the understanding of ensoulment appears to depend on an underlying substance dualism. A closer look, of course, reveals that Pope Benedict XVI does not fully embrace substance dualism. If the soul is not viewed as an immortal or spiritual substance added to an otherwise physical entity at an agreed upon moment in time, then what does this do to the argument for early protectability?

It is our position that the soul is best understood as something relational. Our soul is immortal only because God relates to us in the soul, because God promises to raise us to eternal life. This relationship we dub "spiritual." Without this spiritual relationship with the everlasting God, no such thing as immortality could occur.

Similarly, we assert that dignity has a relational component to it. It is our observation that dignity is relational before it becomes innate. Dignity is first conferred relationally, then it is claimed independently. Where this leads, theologically speaking, is to the observation that dignity is the result of grace. One of our distinguished colleagues at the Graduate Theological Union (GTU) in Berkeley, Roman Catholic moral theologian Richard M. Gula, makes this clear: "As long as God offers divine love (i.e., grace), humans will ever remain God's image and enjoy a sacred dignity whether in sin or not, whether acting humanly or not." This dignity is the gift to us from God's love. It is reinforced by saying we can do nothing to warrant it or merit it. "Human dignity does not depend ultimately on human achievements, but on divine love."[12] Dignity does not depend on a metaphysical substance imparted to the zygote; rather, it depends upon God's call to be the beneficiary of divine love and our love.

Key to the ethical arguments within all three frameworks—embryo protection, human protection, and future wholeness—is the ultimate respect we show to persons through our concern for their human dignity. Each human being is precious, valuable at incalculable cost. The human person is the end or goal of all moral deliberation, never merely a means to some end of greater value. The incompatibility of the three frameworks with each other is determined by asking: To whom does dignity apply? For the *embryo protectionist*, the early embryo from conception on possesses dignity, and we need to protect the life of the early embryo from sacrifice to a further scientific or medical end. For the *human protectionist*, dignity applies to a generic human nature, to finitude and family and to life's meaning found in appreciating our natural state; and the fear here is that the dignity of natural life will become obscured through unguided and chaotic genetic technology. For the *future wholeness* framework, dignity is imputed to the countless millions of persons suffering from physical maladies such as heart disease, diabetes, Parkinson's, Alzheimer's, and cancer; these are persons whom medical science could serve with vigorous research into the therapeutic use of human embryonic stem cells. Stem cell research confers dignity on these future patients.

The biblical mandate to love one another means, among other things, the imputing of dignity to persons, treating them as having worth, value. As we see it, our contemporary ethical mandate is this: We should confer dignity on human persons suffering from disease and trauma for which regenerative medicine might be able to provide relief or even a cure. In large part, public support for the rapid advance of research into regenerative medicine is one important way the present generation could confer dignity by leading so many among us from suffering into abundant life.

Notes

1. Vatican: International Theological Commission, Communion and Stewardship: Human Persons Created in the Image of God 2002 (30), www.vatican.ca/roman_curia/congregations/cfaith/cti_documents/rc_con_cfaith_doc.

2. "What experience explains my having the belief that I am a soul?" asks Stewart Goetz. "The most plausible answer is that the experience in question is just my inner or introspective awareness of myself as a simple substance that exemplifies psychological properties." Goetz, "Substance Dualism," in *In Search of the Soul: Four Views of the Mind-Body Problem*, ed. Joel B. Green and Stuart L. Palmer (Downers Grove,

IL: InterVarsity Press, 2005), 39. For a more comprehensive analysis of the variety of positions on the soul, see Ted Peters, *Anticipating Omega: Science, Faith, and Our Ultimate Future* (Göttingen: Vandenhoeck & Ruprecht, 2007), chapters 6 and 7.

3. Shankara, *Self-Knowledge (Atmabodha)*, tr. Swami Nikhilananda (New York: Ramakrishna-vivekananda Center, 1980), 88.

4. Early Christians rejected reincarnation for a second reason, namely, in the resurrection it is our former body that becomes transformed and reconnects with the soul. In "On the Soul and the Resurrection," Gregory of Nyssa contrasts what the Christian holds from what the reincarnationist holds this way. "The divergence lies in this: we assert that the same body again as before, composed of the same atoms, is compacted around the soul; they suppose that the soul alights on other bodies."

5. The term *creationism* has two quite different meanings in contemporary theology. Among Evangelicals, *creationism* refers to the school of thought that is vigorously opposed to Darwinian evolution. Biblical creationists and scientific creationists argue that God created all species as fixed in their kinds, implying that natural selection will not explain the origin of species. This view developed only in the twentieth century. When we speak of the position advocated by the Vatican in bioethics, *creationism* refers to the ancient view that God creates an individual soul for an individual person. The traditional use of this term does not specify when God creates our soul, only that the soul is not inherited from either eternity or our parents.

6. Johann Auer and Joseph Ratzinger, *Eschatology: Death and Eternal Life*, vol. 9 of *Dogmatic Theology* (Washington, DC: Catholic University of America Press, 1988), 151

7. Auer and Ratzinger, *Eschatology*, 106. The pontiff further recommends that moral theologians engage new problems with creative deliberation. "The moral theologian will also take up the new questions that new developments and relationships pose for the traditional norms." Pope Benedict XVI, *On Conscience* (San Francisco, CA: Ignatius Press, 2007), 72. Perhaps the stem cell controversy provides one of the new situations requiring a new assessment of substance dualism.

8. "In the OT it never means the immortal soul, but is essentially the life principle, or the living being, or the self as the subject of appetite and emotion, occasionally of volition." N. W. Porteous, "Soul," in *The Interpreter's Dictionary of the Bible*, ed. George Arthur Buttrick, 4 vols. (Nashville, TN: Abingdon Press, 1962), 4:428.

9. Goetz, "Substance Dualism," 39.

10. John R. Searle at the University of California at Berkeley provides an example of rejecting religion as scientifically out of date. "Nowadays, as far as I can tell, no one believes in the existence of immortal spiritual substances except on religious grounds. To my knowledge, there are no purely philosophical or scientific motivations for accepting the existence of immortal mental substances." Searle, *The Rediscovery of the Mind* (Cambridge, MA: MIT Press, 1992), 27. For Searle, mind and consciousness are natural phenomena based in our brain biology. Even though neuroscience has not yet proved this to be the case, Searle has faith that someday it will. Searle is critical of substance dualism but admits that a philosopher is able to provide a weakening but

not yet a full destruction of the view. Substance dualism is weakened because (1) no one has ever succeeded in giving an intelligible account of the relationship between the two substances, body and mind; (2) it is possible to account for mental activity within the natural domain, making dualism unnecessary; and (3) substance dualism makes it impossible to show how mental states can cause physical events. Searle, *Mind* (Oxford, UK: Oxford University Press, 2004), 132.

11. The soul is our "intellectual essence," wrote Gregory of Nyssa. This is key to our relationship with God, he thought. "The speculative and critical faculty is the property of the soul's godlike part; for it is by these that we grasp the Deity also." Gregory of Nyssa, "On the Soul and the Resurrection" in *Nicene and Post-Nicene Fathers of the Christian Church*, ed. Philip Schaff (Buffalo, NY: The Christian Literature Company, 1898 and Grand Rapids, MI: Wm. B. Eerdmans, 1994), second series: v: 433.

12. Richard M. Gula, S.S., *Reason Informed by Faith* (New York: Paulist Press, 1988), 64–65.

CHAPTER EIGHTEEN

~

The Ethics of the Ethicists

Ethics is dangerous! If you want to avoid trouble, avoid ethics. Albert Jonsen, one of the most distinguished of the first generation of bioethicists, has written: "Although genetics is beautiful as science, it is an ethical minefield."[1] He is right, as we have discovered.

So, then, why do we do it? Why do ethicists who have perfectly good academic jobs take on more work, such as consulting in the ethical minefield of stem cell research? The work is arduous, the outcome uncertain, the rewards mixed. Indeed, ethicists are under attack when they do consulting work. Until we cut our ties with Geron in 2002, each time one of us would give a public lecture on stem cell research, we could expect to be asked: how much do you get paid by Geron?! The unsaid attack went like this: "You support this research because if you didn't, you'd get fired from the Ethics Advisory Board. You are nothing but a lackey for the corporation." Or, "ethicists who get paid for being ethicists cannot be trusted because their advice is tainted with filthy lucre."

So, this chapter is about the frustrations and joys and the ultimate motivation for our work as ethicists. Why do we do it? Where do we experience frustrations? Is it worth it? Our motivations and experiences differ, but there are some common themes. We begin with the frustrations.

The First Frustration: Time

The first and most obvious frustration of the Geron Ethics Advisory Board (EAB) was time constraints. As indicated in earlier chapters, the announcement of the isolation of human embryonic stem cells put severe time constraints on the work of the EAB. We did not have time to mull things over at leisure.

Here is an example. The guidelines of the EAB include the requirement that "women/couples" should give informed consent. Should it be women? Or couples? What if the woman wants to donate blastocysts, but her partner does not? This is a very complicated issue that would require many lengthy discussions. We did not have time for such discussions if we were to meet our deadlines and publish our recommendations. Hence, we settled for the rather vague phrase "women/couples." Much more precise ethical work remains to be done.

The Pace and Direction of Science

A second frustration was the gap between the pace of science and the pace of ethics. Science moves very fast and it continually changes direction depending on how research unfolds. Every time the EAB met, the science was different than it had been the time before. As one member put it, we sometimes felt that we were "panting to keep up." At one point, we scrapped two years of hard work on issues of histocompatibility. We had analyzed six proposed options for overcoming immune rejection. Then, suddenly, the science changed and washed out our well-thought-through deliberations. We were out of date before we could even write up our analysis. As a dialogic and deliberative enterprise, ethics requires time for discussion, time for disagreement, time for reflection, and time for consultation. But time for deliberation is precisely what may be absent when dealing with fast-moving scientific discoveries.

In our preface we suggested that the image of fast-moving science being trailed by slow-moving ethics is distorted. Here, we admit that the image is partially accurate. Stem cell science is racing down the freeway above the speed limit, while ethicists are still shifting into fourth gear. Still, our important observation is that ethicists are also racing down the freeway. We work hard to keep up with scientific changes and discoveries.

Confidentiality

A third frustration was confidentiality. Often, we learned what was happening in the laboratory months before it could be made public. As the first chapter makes clear, we knew about this research long before the general public did. When Françoise Baylis charged that the Geron EAB signed off on guidelines three years after Thomson's research was funded by Geron,[2] she is partly correct: our official guidelines were published concurrently with the announcement of Thomson's isolation of embryonic stem cells. But what Baylis did not know is that we had been at work on the underlying ethical questions for two years prior to the announcement. During the time that we were consulting for Geron, people would ask questions—or make assumptions that we knew were wrong—and we could not respond because we were operating under a contract that required that we keep confidence on what we knew about the developing science.[3] For those of us who are used to an academic atmosphere of "freedom of speech" and wide sharing of information, this was difficult. It was particularly difficult not to be able to confide in other colleagues or seek their wisdom on issues. We were grateful to have at least the other members of the EAB to check our perceptions and arguments.

Giving up Other Pursuits

Working on stem cell ethics required considerable investment of time and energy. It derailed some of our other work.[4] Stem cells took over our professional lives for nearly four years and left little room for other commitments. Although we were excited by the stem cell work, it was frustrating to give up scholarly projects into which we had previously invested considerable time and energy.

Public Attacks

A fifth frustration had to do with finding ourselves in the limelight and receiving public criticism of our work. At times, others would misrepresent the task and the mandate of the EAB. As noted above, on occasion we were publicly criticized. Our integrity was impugned. At the time, we could not respond. Because we were not a public organization but a group of consultants contracted to offer advice *to the corporation*, we had no obvious public forum in which to respond to criticisms. Nor would it have been appropriate for us to spend our time responding to criticisms rather than doing the work for which we had been engaged.

One of the milder criticisms is that we were incompetent. We were dubbed incompetent because we were theologians and ethicists without scientific or legal membership on the EAB proper. This criticism we find curious, in light of our constant work with scientists from Geron. What we brought was expertise in ethics that would otherwise be lacking in a corporation focused on scientific research. The very fact that we were theological ethicists ironically brought a second criticism—that as theologians and ethicists, we were too narrow in our expertise. Cynthia Cohen said, "The Geron board is also open to possible criticism on grounds that its members were all from the same field and had insufficient expertise in areas related to stem cell science and in addressing the concerns of patients and the public." Fortunately, Cohen adds to this criticism a backhanded compliment: "Yet this board issued a set of guidelines and presented an analysis of the issues that reflected an understanding of the science involved. . . . It went on to reach conclusions for which it provided reasonable grounds."[5] Despite our alleged incompetence, we were still able to reach conclusions based upon reasonable grounds. Imagine that!

We have taken to heart sharper criticisms raised by bioethicist Ron Green, a consultant to Michael D. West at Advanced Cell Technology (ACT) in Boston. Green's concern is that federal funding should have been given to stem cell research. But, due to the Dickey policy that forbade federal support for embryo research, private biotech companies such as Geron stepped into the vacuum and became dominant. He faults the U.S. government for its shortsightedness. By implication, he suggests that bioethicists in the employ of the private sector are less than fully objective in their assessments.

He speaks about our Geron EAB directly: "the limits of Geron's in house EAB. . . . The advisory board's guidelines seemed to have been formulated *after* Gearhart and Thomson conducted the bulk of their research. Just as the embryo panel [U.S. Human Embryo Research Panel] had predicted . . . the opportunity to conduct stem cell research and the government's forced abandonment of funding in this area had given rise to an ambitious private sector research effort the ethical review of which raised questions."[6] Once again, the assumption is that we were brought in at the last minute to "rubber stamp" the science. Once again, we would point out that we were there from the beginning.

Green is not alone. We were frequently maligned because Geron paid us for our work. Some who opposed stem cell research would rhetorically sweep up scientists and ethicists into the same garbage bin. The U.S. House of Representatives heard this in a speech on the floor: "Mad scientists are still mad scientists no matter how white their lab coats are and how many bioethicists they have to justify their actions."[7] The chair of the U.S. President's Council on Bioethics wrote this: Bioethicists have entered "in large numbers into the employ of the biotechnology companies bestowing their moral blessings on the latest innovation—assuredly not for love but for money. If these 'experts' can't see or don't care about what lies ahead, what hope is there for the rest of us?"[8]

The very question about pay is an implicit attack on the integrity of the ethicist's work, and hence on the integrity of the ethicist. Rather than interview us and ask us about what were doing, these public spokespersons felt they could simply tarnish our reputations by associating consultant fees with selling our ethical souls to immoral science. This is enough to give an ethicist a bad day!

So, we needed to think about it. And we did. Two concerns obtained. First, Geron asked—even demanded—that we retain independent judgment. We would be of no value to the corporation if we were to simply copy party line. We were not asked to advertise or promote Geron products, or to appear as poster ethicists to enhance the company's public image. Indeed, we were specifically enjoined against owning any corporation stock. We could be of value only as a group of professionals behind the scenes rendering independent ethical evaluations and judgments.

The second concern was receipt of the consulting fee for our work. We debated whether or not to work for compensation, and we decided to accept such payment for our time and energy. Such payment would be appropriate—indeed, expected—in all other comparable professional situations. It is common for academic people to work for both their host institution as well as the wider scholarly world. Delivering lectures, conducting workshops, publishing, and doing consulting work are common in academia, and they are commonly compensated. Many bioethicists on faculties of universities also work as hospital consultants. They are paid for this work. Does being paid mean that they will automatically operate with the *hospital's*, rather than the *patient's*, best interests in mind? Others are invited to give lectures—in essence, to express their opinions—and are paid for doing so. Why, then, should it be wrong for ethicists to be paid by corporations?

Further, the demands on our time were for many periods considerable, and it is entirely appropriate to be paid for one's time. Hence, our response to the question of pay is simple: "Of course we get paid. Our love is free; our time isn't." In fact, however, the honorarium paid to the EAB was modest[9] and covered only the days we were together in formal meetings. It did not include pay for time spent doing research behind the scenes, or for our time spent writing. In addition, members of the EAB have consulted with CEO Tom Okarma of Geron on a number of informal occasions for which we have not been paid. It is frustrating, however, to have one's integrity constantly impugned by allusion to dollar signs.

A Diversity of Ethical Perspectives

Finally, collaborative decision-making also has its frustrations. As noted earlier, we tried to find common ground to support positions taken. But some colleagues were much more utilitarian than others, and we did not always agree on the fundamental orientation we should take. Diversity ensures that there is no single party line, but it also makes consensus difficult. In consultation with the CEO of Geron, we expressed a range of opinions; but in order to write for public purposes, we had to agree on language to express our ideas. Sometimes, we had to compromise—for instance, agreeing to a "developmental" view of the moral status of the embryo, when each of us might have meant something different by that term. Once a compromise is on paper, we had to defend it even if we had disagreed with it originally.

At the same time, we asked ourselves: Was our diversity sufficient? Our group included Jewish, Roman Catholic, and several branches of Protestant backgrounds. But we had no Muslim or Buddhist members. Could we then speak to a diverse population such as that of the United States? In order to work well together, we had to be small. Being small meant, however, that we risked ignoring views from other traditions. That we did attend to those traditions is evidenced by our efforts here to speak to several of them.

Another frustration of collaborative work is the "least common denominator" phenomenon. "Ethics by committee" forces us to challenge and clarify our own positions but tends in the long run to depend on well-established concepts. In much of our work, we drew on the ethical principles that have become enshrined in the public literature on bioethics: respect for persons, beneficence, non-maleficence, and justice. Because these principles are well established, they provide a firm foundation for work in the field. But they are not strikingly new. Ethics often works by small incremental advances over previous thought. Science works this way, too. Stem cell work of various sorts had been going on for many years prior to the breakthroughs of 1998. However, those breakthroughs were so dramatic that they seemed not like incremental advances, but like sea changes. Some people expected the ethical reflection also to be a sea change, not a repetition of old principles. Did we say anything significantly new? Have our guidelines and pronouncements been enduring and important? To the extent that collaborative work requires dependence on broad principles that are widely accepted, it may not be stunningly creative.

Furthermore, for those of us who work out of explicitly theological foundations, the use of principles framed in nontheological language can be frustrating. Beneficence and non-maleficence are principles that now communicate to a wide audience. However, they do not capture the subtleties that inhere in specifically religious concepts such as the Roman Catholic concept of "double effect" or the Protestant notion that humans have not only a "nature" that must be respected, but also a "destiny" that determines what we should do. The "principles" approach of contemporary philosophical ethics is contrary to Jewish halachic reasoning and tradition. Bridging the gap between theology and secular discourse carries its own frustrations.

The Joys

Given all of these frustrations, one might ask why we would bother. Why do it? Why derail other projects, engage in the compromises of collaborative work, subject ourselves to uncomfortable time constraints and confidentiality requirements, open ourselves to public attack, and struggle with the gap between theological and philosophical language? What is the reward and motivation that makes us embrace this work?

First and foremost, we did this because it was a public service. From the first meeting with Michael D. West and our growing understanding of the potential significance of stem cells for human health and well-being, we knew that the science in this project would be huge and globally important. We anticipated that it would have ethical implications requiring serious consideration; and we believed that we would be in a unique position to provide that ethical reflection. While it is true that we were constrained by confidentiality as to the specifics of Geron's work, it is also true that we were able to begin framing issues long before they hit the public arena. As indicated in our book's early chapters, our framing of those issues included a number of concerns that have only recently been picked up more generally. We were on the forefront of ethics, and we forecasted that our work could be significant.

We had had the privilege of being involved in other settings where we could influence public policy.[10] When we joined the Geron EAB, we did not know whether or how quickly we would get thrust into the limelight, but we were glad to have opportunities to testify before the National Bioethics Advisory Commission and to consult with the American Academy for the Advancement of Science. Thus, not only did we do this as a public service, but we had reason to think that our work was making a difference. This conviction kept us going during some otherwise frustrating times.

One of the places where we know our work was making a difference was in the corporation itself. Near the end of the first year of the work of the EAB, Geron was considering a research protocol that would involve embryos. As we reviewed this protocol, some members of the EAB objected. We discussed the issues and implications at length, generating considerable heat. Although we never took an official or unanimous position on that research, the CEO decided not to support that research direction, based on the concerns raised by the EAB. Six months later the protocol was re-written and resubmitted to the EAB. It passed the EAB, and the research commenced. Our ethical instincts were honored and our perspectives re-spected. To those who think we were only lackeys of the corporation, we can confidently reply that we caused the corporation to change direction on at least one significant occasion.

Of course, our work was not always "line drawing" or "nay-saying." One of the joys of partnering with scientists was having the opportunity to support research that we consider valuable. Because the embryo protectionist frame-work in its negative interpretation seemed to get a grip on the public discus-sion, we worked with a sense of urgency to provide arguments supporting the research within the future wholeness framework. Sadly, the public noise sur-rounding stem cells has been so dominated by voices claiming that embryo protection is "the" Christian position that our supportive deliberations have been nearly drowned out. That is one reason for writing this book: We want the world to know that some theologians *do* support stem cell research!

The Ultimate Motivation: Faith

But for the three of us who wrote this book, there was yet one more motiva-tor, and it was the important one: faith. A century ago a German theologian, Ernst Troeltsch, connected ethics with newness. If theological principles are to be helpful in a rapidly changing society, wrote Troeltsch, "thoughts will be necessary which have not yet been thought, and which will correspond to this new situation as the older forms met the need of the social situation of earlier ages."[11] We see the task of the theological ethicist as providing creative guidance in the middle of new situations. The new situations will just keep coming at increasing speed, so we predict that ethicists will earn an honest living for some time to come. In the case of stem cell research, some ethicists need to hold up the vision of human flowering through the relief of suffering and living a healthier and longer life. We believe this to be a fundamental ethical concern, one that comes to articulation within the future wholeness framework.

It is our faith that leads us to public service, even at some personal cost. It is our faith that tells us to honor truth, in the conviction that "truth shall set you free." It is our faith above all that has led us to stress from the very beginning three concepts that make a difference in thinking about stem cell research.

First, the dynamic relation between nature and destiny. We believe that our inherited created nature is only part of the ethical framing for our life on earth. The other part is our future or destiny as God's children.[12] Rather than look for something sacred in our past or in our genes or in our cells, we look forward to the promise of God's transformation of what we have inherited. We think of our *nature* in creative tension with our *destiny*. John 3:16 tells us that this destiny is "abundant" life. This passage is sometimes translated "eternal life." Such a translation invites the idea that we are looking to a life after death. However, as "abundant" life, the phrase looks not merely to life after death but to the quality of life here on earth. Because the Bible promises abundant life in eternity, we believe it is our ethical responsibility to pursue the abundant life here on earth in anticipation of God's intended destiny. It is this call to abundant life that orients our support for stem cell research with its possibilities for "regenerative" medicine.

Second, justice. Our faith makes central for us the concept of *justice* in the ethical analysis of stem cell research. Even before stem cells became a public topic, we had begun to talk about the importance of justice in addressing this research. A theological concept of justice may be a bit different than the framing of this concept in the secular arena.[13] Justice in philosophical and secular contexts is often taken to mean "equal opportunity" or "similar cases should be treated similarly." For those of us from biblical traditions, however, justice has both a more overarching meaning and a more specific demand. Its overarching meaning is that everything must be in "right relationship." Equal opportunity or treating similar cases similarly are not broad enough concepts to encompass this sense of overarching justice or "righteousness." The specific demand that accompanies justice as righteousness is a preferential option for the disadvantaged—the poor and oppressed. Thus, from the beginning, we have been concerned to bring the discussion of stem cell ethics into the public arena where questions of justice can be widely addressed, rather than leaving that discussion to the privately funded scientists pursuing the work of stem cell research.[14]

What justice means to us at this stage is this: All involved parties—scientists, universities, funders, biotech companies, investors, insurance companies, faith communities, and regulatory agencies—should conspire together to ensure that the products of this expensive research be made available to every human being around the world, including the poorest of the poor. Nothing less than universal access to the benefits of exotic science belongs in a vision of justice.

Third, salvation as health and wholeness. Our faith cannot help but express itself as concern for good health combined with a confident attitude in the face of what is less than good health. The Christian word *salvation* is tied to words in various languages meaning "health." Jesus was a healer as well as a savior. The promise of eternal salvation is anticipated by our prayers for good health. Enjoying good health now anticipates future fulfillment, and both are included in the "abundant life." To support stem cell research in hope of bettering human life and expanding human well-being is a secular form of holy work; it gives expression to God's plan for healing. Science may fail, of course; the promises of stem cell research may fail to come to pass. Yet, it is only good stewardship to act on these promises while the opportunity presents itself.

Even so, the relationship of faith to healing is not solely dependent on the latest scientific research. For a person of faith, the concept of health picks up many subtle nuances. Health is not merely the absence of malfunctioning in our bodies. Health has physical, psychic, relational, and spiritual dimensions. True health is found in a kind of strength described by Reformed theologian Jürgen Moltmann: "But if we understand health as the strength to be human, then we make being human more important than the state of being healthy. Health is not the meaning of human life. On the contrary, a person has to prove the meaning he has found in his own life in conditions of health *and* sickness. Only what can stand up to both health *and* sickness, and ultimately to living *and* dying, can count as a valid definition of what it means to be human."[15]

On the one hand, health is physical well-being. On the other hand, health is the soul's inner attitude of trust and confidence that sustains our dignified humanity in the face of bodily limitations or sufferings. Such confidence can be inspired by the promise of eschatological fulfillment, by God's promise that in the New Jerusalem we will be liberated from crying and pain. God "will wipe every tear from their eyes. Death will be no more; mourning and crying and pain will be no more, for the first things have passed away" (NRS, Revelation 21:4).

Balancing the Frustrations and Joys

Now, we do not want to present ourselves as more virtuous than we really are. Being good Augustinians, we readily admit to our flaws and failures. We did have our interests too. As academic scholars, we love to be on the "cutting edge" of new developments. Whether it is pioneering a new method in one's field or being the first to research a new topic or offer a new perspective, one of the constant motivations for thinkers is thinking new thoughts. Our work as consultants to a corporation pioneering in the science of stem cells also gave us a chance to be on the cutting edge of both the science and the ethics of stem cells. For a number of years, we probably knew more about what was happening in stem cell research than did any other group of nonscientists. It was exciting to be on the forefront of the development of new knowledge. Our motive was largely to serve, but it had selfish edges, too.

Finally, having noted above some of the frustrations of collaborative work, we would add that the joys outweighed the frustrations. Having to defend a position to knowledgeable and thoughtful colleagues helped each of us to sharpen our own views. Having to compromise taught us the value of some concepts that did not come "naturally" to our own traditions. Some of our meetings were contentious. We had to work through differences in style, language, and approach. But over time, we came to understand and appreciate each other's views. If one of us could not attend a meeting, someone else would speak up and say, "If so-and-so were here, she or he would say. . . ." We could almost anticipate each other's reactions and common themes.[16] To this day, we retain a deep respect for each other's thoughts, even when we do not fully agree. We would not be writing this book collaboratively if we did not find value in collaborative work.

Thus, the frustrations of the work were more than balanced by the joys—the sense of public contribution, the joy of being involved in forming policies for guiding significant scientific work, the pleasure of learning with good colleagues, the excitement of being on the cutting edge of new developments in science and ethics, and the sense that we were living out what our faith would lead us to do.

Notes

1. Al Jonsen, *The Birth of Bioethics* (Oxford, UK: Oxford University Press, 1998), 167.

2. Elaine Deward, *The Second Tree: Stem Cells, Clones, Chimeras, and Quests for Immortality* (New York: Carroll and Graf Publishers, 2004), 416.

3. For instance, Eric Parens of the Hastings Center called Karen Lebacqz early in 1999 with a number of questions about the implications of stem cells for cloning. In fact, Lebacqz was preparing her own statements on that very topic, but all she could say was, "You know that I am not at liberty to discuss the work of the EAB."

4. At the time when we began our work, Karen Lebacqz was due a short sabbatic leave, on which she intended to write a book on professional ethics for clergy. The book was never written!

5. Cynthia B. Cohen, *Renewing the Stuff of Life: Stem Cells, Ethics, and Public Policy* (Oxford, UK: Oxford University Press, 2007), 216.

6. Ronald M. Green, *The Human Embryo Research Debate* (Oxford, UK: Oxford University Press, 2001), 134.

7. U.S. Congress Representative (R-N.J.) Chris Smith, cited in Michael D. West, *The Immortal Cell: One Scientist's Quest to Solve the Mystery of Human Aging* (New York: Random House/Doubleday, 2003), 160

8. Leon R. Kass, *Life, Liberty and the Defense of Dignity: The Challenge for Bioethics* (San Francisco, CA: Encounter Books, 2002), 10

9. Our honorarium was deliberately modest, precisely so that it would be clear that we were not biased by financial considerations.

10. For instance, Ted Peters was principal investigator for an ELSI grant in conjunction with the Human Genome Project; Karen Lebacqz was a member of the National Commission for the Protection of Human Subjects of Biomedical and Behavioral Research, which proposed guidelines for research on human subjects that were turned into government regulations.

11. Ernst Troeltsch, *The Social Teaching of the Christian Churches*, 2 vols. (New York: Harper, 1931), 2:1012.

12. See Ted Peters, *Futures: Human and Divine* (Louisville, KY: Westminster/John Knox Press, 1977).

13. For further discussion of the demands of justice and the differences between philosophical and Christian perspectives, see Karen Lebacqz, "Justice," in *Christian Ethics*, ed. Bernard Hoose (Herndon, VA: Cassell, 1998). Lebacqz is also the author of two books on justice: *Six Theories of Justice* (Minneapolis, MN: Augsburg, 1986) and *Justice in an Unjust World* (Minneapolis, MN: Augsburg, 1987).

14. To say this, however, is not to say that those scientists are motivated only by financial gain. Indeed, quite the opposite is true. Roger Pederson, John Gearhart, and James Thomson were all motivated into stem cell research, so far as we know, by experiences of disability among friends or family. Thus, their work is itself an expression of a commitment to the "poor and oppressed" among us.

15. Jürgen Moltmann, *God in Creation* (San Francisco, CA: Harper, 1985), 273.

16. For instance, our colleague Ernle W. D. Young constantly reminded us of the bounty and superfluity of nature. His arguments for the acceptability of embryo loss depended in part on his understanding of how much loss is a normal part of natural processes.

CHAPTER NINETEEN

~

Theologians Say "Yes" to Regenerative Medicine

The now 10-year-old prophecies that human embryonic stem (hES) cell research would revolutionize medicine have not yet come to pass. Crucial advances toward therapeutic proof-of-concept have been made, to be sure. However, differentiation is still not understood well enough to be managed. Stem cell transplants still result in complications more frequently than is safe for regular experimentation in human subjects. And the problem of immune rejection, and its connections to nuclear transfer, remains a blockage point. Tie-ups on the political front are often cited for why the pace toward therapeutic revolution has been slow. Regardless of whether it is due to slow science or recalcitrant politics, the fact remains that as of the writing of this book, the revolution has not come. Regenerative medicine remains a promising and plausible idea, but an idea whose applications have not born benefits equal to the prognostications.

Despite the slow pace toward therapeutic application, hES cell research has had a notorious influence on social and ethical thinking. This is most obvious in the ways in which stem cell research has impacted U.S. politics. In the last half-decade stem cell research has been given a place alongside abortion and evolution as a litmus test of one's political, moral, and religious affiliations. Stem cell research has proven influential for a second, less publicly visible reason as well. Stem cell research has impacted our understanding of how nature works. It has given us new insight into the nature of nature, demonstrating that biological systems are much more flexible and malleable than previously thought.

In short, stem cell research has changed our thinking about nature's potentials. This means that stem cell research has changed the way in which problems are taken up in all three religiously grounded frameworks. New understandings of biology change the way in which we think about the early embryo and potentiality. It changes the way in which we think about human nature. And it changes our thinking about the potential of medical research to contribute to an abundant human future.

Of course, stem cells will not make us immortal, nor will stem cell research alone give us an abundant human future. Yet, we as Christian theologians say "yes" to stem cell research and the development of regenerative medicine. On this side of our future wholeness, God calls us to life, and to life more abundantly. The noble practice of medicine is a human practice that carries out a divine mission on earth. Doctors and nurses, along with the scientists who further their skills, help to redeem us from suffering and, thereby, contribute to human flourishing. Medical science need not deliver immortality to participate in God's saving work among us.

In what follows, we would like to provide a summary of our deliberations regarding regenerative medicine from within the future wholeness framework. We would like to explain just why we say "yes" to stem cells. We offer these arguments based in Christian theology, fully recognizing that what we say must be placed before the bar of ethical reasoning within the larger public sphere with its pluralism of religious and moral perspectives. In what follows we offer seven items, each leading to a cautious yet enthusiastic theological "yes" to stem cell research.

1. Genuine Immortality

Will stem cell therapy make us as persons immortal? No. Even if pluripotent stem cells can be called "immortal," it does not follow that persons with regenerative stem cells in their bodies will become immortal.

Indeed, we could be misled by stem cell hype. One Egyptian god associated with the immortality of the pharaohs, Osiris, would be surprised to find his name associated with a mesenchymal stem company in Baltimore. Isis, Osiris's wife, responsible for Osiris's resurrection from drowning in the Nile, finds her name in a company called Isis Pharmaceuticals. Cynthia Fox interprets these symbols. "In choosing the names of Egyptian gods to market their cells, researchers have reached over the colossus of modern science to a time of colossal faith, when man was so certain about the prospect of eternal life he built monumental steps to it," namely, the pyramids.[1]

When Christian theologians speak of eternal life, do they refer to extended bodily life due to good health? No. The biblical promise for eternal life is connected to doctrines of resurrection from the dead and the advent of the new creation. The biblical promise is that God will transform and renew present reality so that "death will be no more" (Revelation 21:4). The only door to this new creation is the cross followed by Easter, death followed by resurrection. Acceptance of our death is built into this understanding of immortality. "Christ has been raised from the dead, the first fruits of those who have died" (1 Corinthians 15:20). As Christ was raised on Easter, so too will we be raised at the dawn of God's new creation. This is what eternal life refers to in the Christian vocabulary, and living today can be made more abundant if we are inspired by this future vision of what is to come.

As much as we support regenerative medicine, we want to avoid any confusion between what marvels come from science and what marvels are promised by God. We place our faith in God, not in science. No amount of enthusiasm for stem cells should permit a lack of realism regarding what might be accomplished here. We wish to celebrate the more abundant life that regenerative medicine might bring, but we also wish to avoid the disappointment of unrealizable expectations.

2. Beneficence and the Vision of God's Future

The scientific context within which we find ourselves is one in which pluripotent hES cells have been isolated and characterized. With concentrated laboratory effort, it is possible that these pluripotent cells can be trained to become any tissue we target. Once integrated into a human organ such as the brain or heart or pancreas or liver, these cells could regenerate tissue and restore organ function. The therapeutic value lies beyond our imaginations. Untold yet is the possibility of repelling degeneration from Alzheimer's, Parkinson's, heart disease, spinal cord paralysis, diabetes, and perhaps even cancer. With this vision of a potential future, just what is our moral obligation? We believe our society should fertilize the seeds of this science and plan for a wide and equitable distribution of its eventual fruits.

Given that Hippocrites said, "benefit, and do no harm," we give primacy to "benefit." Beneficence is the ethical principle for seeking benefit. If we believe the nature we have inherited is transformable, we will seek levels of health and well-being beyond what we have inherited from our evolutionary and biological past. What determines our ethical vision is not what seems fixed by the past, but the potential that lies in the future.

Regenerative medicine has the potential for delivering enormous relief to human suffering and for enhancing human flourishing. Our moral responsibility, we believe, is to seek to actualize this potential. It is the vision of healing that draws us forward. As Christian theologians we are compelled by the vision of the final book of the biblical corpus, Revelation. Here we find the new Jerusalem descending from heaven. Citizens in this eschatological city will discover that they are healed. No more will there be crying or pain. "Then I saw a new heaven and a new earth; for the first heaven and the first earth had passed away, and the sea was no more.[2] And I saw the holy city, the new Jerusalem, coming down out of heaven from God, prepared as a bride adorned for her husband.[3] And I heard a loud voice from the throne saying, 'See, the home of God is among mortals. He will dwell with them; they will be his peoples, and God himself will be with them;[4] he will wipe every tear from their eyes. Death will be no more; mourning and crying and pain will be no more, for the first things have passed away'" (NRS, Revelation 21:1–4).

From within the future wholeness framework, we begin our ethical deliberation from a vision of healing, a vision of life in abundance as God has promised it to us. To take advantage of an opportunity provided by contemporary science is to align our social responsibility with the future kingdom of God.

3. Beneficence Adds to Non-maleficence

For those arguing from within the embryo protection and human protection frameworks, the avoidance of harm seems to suffice for ethical reasoning. As long as we do not disturb the natural processes we have inherited from either God or nature, we will have fulfilled our moral responsibility. As long as we assume the *ex vivo* blastocyst is ensouled, sacred, and possessing morally protectable dignity, we will have performed our ethical task. As long as we protect the early embryo from destruction in the laboratory, we can claim that we have fulfilled our moral obligations. As long as we avoid doing harm, we can exonerate ourselves. As long as we do nothing, we can claim to be in the right.

We find this to be an inadequate approach to ethical reasoning in the present context. In the present context, science stands up to announce that a better world is possible. With some concentration of energy and resources we can likely enhance human health and well-being; we can spark a giant leap forward in human flourishing. For some, standing still and invoking non-maleficence suffices. For us, this stops short of the kind of ethical discernment the present situation requires.

We turn to the parable of the Good Samaritan at this juncture. According to the New Testament, Jesus is asked by a man, "What must I do to inherit eternal life?" The man wants immortality, life in abundance. What follows is a dialogue regarding God's law. The questioner claims that he has fulfilled the law, to the letter. Yet, something is missing. What is missing is compassion for the neighbor, an aggressive beneficence. In order to get at what is missing, Jesus tells a story. This story is instructive for us here.

NRS, Luke 10:30–37: "A man was going down from Jerusalem to Jericho, and fell into the hands of robbers, who stripped him, beat him, and went away, leaving him half dead (Luke 10:31). Now by chance a priest was going down that road; and when he saw him, he passed by on the other side (Luke 10:32). So likewise a Levite, when he came to the place and saw him, passed by on the other side (Luke 10:33). But a Samaritan while traveling came near him; and when he saw him, he was moved with pity (Luke 10:34). He went to him and bandaged his wounds, having poured oil and wine on them. Then he put him on his own animal, brought him to an inn, and took care of him (Luke 10:35). The next day he took out two denarii, gave them to the innkeeper, and said, 'Take care of him; and when I come back, I will repay you whatever more you spend' (Luke 10:36) Which of these three, do you think, was a neighbor to the man who fell into the hands of the robbers?" (Luke 10:37). He said, "The one who showed him mercy." Jesus said to him, "Go and do likewise."

We note that in this parable both the priest and Levite could pass an ethical test when non-maleficence is the criterion. When they saw the suffering man, they did not walk over and kick him. They did not add to his misery. They were certainly moral in this sense.

Yet, Jesus seems to lure us toward something more. The Samaritan felt "pity," and he acted on his pity. He found a creative way to provide help and sustenance. We ask: Could the Good Samaritan provide a model for our society today? Could regenerative medicine be understood by analogy as the aid the Samaritan offered to the suffering victim?

The very existence in our society of hospitals named after the "Good Samaritan" is testimony that some have heard the call of compassion to go beyond non-maleficence to beneficence. Might we find a way to think like the Good Samaritan as we confront the stem cell controversy?

4. Social Justice as Entailed in Beneficence

Justice belongs to our vision of human wholeness; and justice is an inextricable element in the beneficence agenda. The benefits of exotic science ought not to be limited to only the wealthy. Because of the cost of such expensive science, and because of the need for patenting intellectual property in order to draw financial investment, our society risks losing all the benefits of stem cell science to an economic system that favors the rich among us.

The question of *access* looms large within the future wholeness framework. Wholeness is not just about individual persons, but is about the entire human community. We ask: How can we ensure access to the benefits of regenerative science on the part of the poorest of the poor, regardless of where they live in this world? Smoke screens and sleights of hand on the part of investors and regulators who make big promises now must find that later, when delivery is ready, a system is in place to guarantee widespread access.

Greedy investors and avaricious patent applicants ought not to take refuge in the rhetoric of ethicists such as ourselves who publicly embrace stem cell research. Our ethical arguments are not intended to prop up economic structures that would continue to divide the world's peoples into haves and have-nots. Intrinsic to our argument within the future wholeness framework is that the benefits accruing from this science be widely distributed so that meeting genuine human needs takes precedence over corporate profit. If profit and service can become partners, then we will rejoice. But, we will not relinquish our prophetic vocation that stands up for the responsibility of the wealthy to care for the poor. And medical care for all on earth, including the poor, is front stage in this ethical drama.

5. Attention to the Embryo Protectionist Concerns

Even though we work from within the future wholeness framework, we feel some obligation to respond to the assumptions and commitments of those within the embryo protection framework, especially those who would like to shut down embryonic stem cell research. Their adherence to the principle of non-maleficence toward the blastocyst is admirable. Even though admirable, we believe that their premises are faulty and their moral stance fallacious.

Our investigations show that the Vatican has proffered the most sophisticated ethical reasoning. The Vatican contends that genetic uniqueness is established at conception even *ex vivo* and is somehow connected to an immortal soul and hence to morally protectable dignity. This contention is riddled with too many questionable assumptions. The assumption that a cell—the zygote or even the blastocyst—can be sacred and thereby demand inviolable dignity cannot pass close theological scrutiny. So, when the Vatican and its American evangelical partners fire all their artillery against laboratory researchers in order to shut the science down, we ask whether this is the best among the moral alternatives.

What is unknown even within Catholic circles is exactly when ensoulment occurs and when dignity is established. Yet, Vatican arguments push moral protectability all the way back to conception, so that they are better safe than sorry. All the weight here is placed on non-maleficence. The protection of the embryo trumps all concern for the person who might benefit from research into regenerative medicine.

We believe the better path to follow is to construct a moral argument supporting regenerative medicine from within the ethical framework of future wholeness. What we do in fact know without any doubt is that millions if not billions of human individuals could benefit from therapies developed through stem cell research. Knowledge of actual human beings who suffer is indisputable. To sacrifice a potential benefit for such persons on the basis of questionable metaphysical assumptions regarding *ex vivo* blastocysts seems to us to be inadequate ethical reasoning.

Having said this, we nonetheless believe that an argument can be mounted from within the embryo protection framework on behalf of saying "yes" to stem cell research. This is why we have appealed to the 14-day Rule. When the early embryo is within the mother's body, *in vivo*, both individuality and relationality become the dialectic that establishes potential personhood. We cannot say with absolute or apodictic confidence that the 14-day threshold is decisive; yet, if we listen at all to science, it tells us something relevant and valuable. We conclude that if the Vatican takes seriously its own criteria for ensoulment, then the 14-day threshold should be appealing. This would leave *ex vivo* blastocysts open for scientific investigation and possible service in the betterment of human life on earth.

6. Theology and Public Policy

We recognize that arguments constructed upon distinctively Christian foundations are specific to our own religious tradition. We do not expect Muslims or Jews or Hindus or Buddhists or secularists to accept our foundational principles. We cannot demand adherence to our views on the basis of religious authority, either biblical or ecclesiastical. As believers in God, begin with our beliefs and draw out implications. For our arguments to gain a hearing in the public sphere or to be persuasive, they must be seen as charitable and reasonable.

One of the messages of this book is that public discussion of the stem cell controversy around the world may look secular, but underneath it is quite religious. Religious beliefs underlie most of the international debate. What has already happened is that our pluralistic global community has translated a variety of religious commitments into presumably neutral secular language and then proceeded to debate issues. Religious points of view are getting argued in a public sphere where no one religious tradition holds dogmatic authority.

In this book we offer our theologically considered argument as one among many. We hope the reader will consider it carefully.

7. We Are Not Alone

We claim Francis Collins as an ally in the future wholeness camp. In the United States, Collins could easily be called "Mr. Gene." Known for his triumphs as the molecular biologist who discovered the gene for cystic fibrosis and the gallant leader of the federally funded Human Genome Project, Collins directs the National Center for Human Genome Research at the National Institutes of Health. Not only is he one of the most respected scientists in today's world, he is a committed and outspoken evangelical Christian. Because of the alliance of evangelicals with the Vatican in both the abortion and stem cell controversies, one might predict that Collins would fall in line with the embryo protectionist view described earlier. But this is not the case. Collins strongly favors hES cell research, arguing primarily from within the future wholeness framework on behalf of medical benefits.

Collins frequently announces that Jesus was a healer. Collins is a medical doctor. This means he is a healer as well. The most "immoral" thing we could do, argues Collins, would be to shut down stem cell research and deny to myriads of suffering people the potential healing that regenerative medicine might bring.

The reader may recall earlier discussions of cytoplasmic reprogramming, or to use the term of the U.S. President's Council on Bioethics, "somatic cell dedifferentiation." Collins appeals to the prospect of this development to address the anxieties of his fellow travelers in American evangelicalism. If we could make blastocysts from DNA taken from skin cells, might this calm fears that we are destroying human embryos and thereby committing abortion? What Collins is imagining is dedifferentiating the DNA nucleus of a skin cell, returning it to its quiescent state. Then, one would place this skin cell within an ennucleated egg. The cytoplasm of the egg would trigger embryogenesis. At the blastocyst stage, pluripotent stem cells could be harvested. In effect, this applies to humans the nuclear transfer technique developed in the cloning of Dolly.

"I would argue that the immediate product of a skin cell and an ennucleated egg cell fall short of the moral status of the union of sperm and egg. The former is a creation in the laboratory that does not occur in nature, and is not part of God's plan to create a human individual. The latter is very much God's plan, carried out through the millennia by our own species and many others."[2]

This particular argument is set within the embryo protection framework. It works on the premise that what occurs in the mother's body through sexual fertilization is natural and thereby godly. What occurs by scientific artifice in the laboratory does not warrant the same theological or moral status. Deriving pluripotent stem cells from reprogrammed skin DNA would satisfy the ethical needs of both the embryo protectionist and the future wholeness proponent. And, as a scientific by-product, it could even provide patient-specific DNA to achieve histocompatibility.

Our own ethical logic is similar, but not identical. Even though we applaud Collins's attempt to meet the demands of the strict embryo protectionists among evangelicals and Roman Catholics, we deem this unnecessary. We believe that the *ex vivo* blastocyst in the laboratory petri dish cannot claim the same moral status as a baby growing in a mother's womb. This is our considered judgment based upon our best reasoning, something short of apodictic dogma. Yet, we must proceed on the basis of our best reasoning and give permission to scientists to conduct their experiments on this genetic material.

Like Collins, we believe that it makes a difference to distinguish between a mother's body and a petri dish. Within the mother's body, we believe the 14-day Rule marks a significant threshold: At this point a relationship between the embryo and the mother is established that will nourish the development of an identifiable individual human person. Although in itself this is not morally conclusive, we believe it counts. No such threshold can be reached in a laboratory petri dish. No potential for development into a human person exists *ex vivo*.

Note how our ethical logic differs from that of Collins, even if only slightly. For Collins, what happens when a man and woman conceive the old-fashioned way is natural and, thereby, God's will. What happens in the laboratory is artificial and, thereby, not part of God's "plan" for bringing a new child of God into the world. The product of scientific artifice is not a human or potential human deserving moral protectability. The contrast Collins lifts up is the contrast between what is natural and what is artifice.

The contrast we lift up is the contrast between *in vivo* and *ex vivo*. At 14 days *in vivo*, a relationship between mother and child is established so that an individual human person is now on the way, so to speak. Potentiality in a general fashion existed prior to this threshold, to be sure; but now at 14 days we meet the Vatican criterion for the first time, namely, we have a unique human individual. Yet, who he or she is as an individual appears in utter and total dependence upon the relationship to the mother. This can happen only *in vivo*.

This dialectic between individual and relation cannot occur in the laboratory, *ex vivo*. In no realistic sense can one speak of a blastocyst in the petri dish absent implantation as a potential person. For us the question is not whether it is natural or artificial; rather, the question is whether or not the basic relationship that enables the development of personhood is in place. This only happens *in vivo*.

Regardless of this difference in ethical logic, what we celebrate are the scientific achievements and the potential for improving human health and well-being that are the outcomes of this scientist's electrifying career.

8. "Yes" to Stem Cells

We are not naive. We are well aware of the contingencies, the hype, the exaggerations, the propaganda, the competition between vested interests, the skepticism, and the risks. Any calm and judicious assessment would have to entertain the prospect that all of this will turn out to be but a puff of smoke that will drift away and dissipate. There is no divine guarantee that investment in regenerative research will ever pay off. It could all go bust.

Nonetheless, as we view alternative futures and alternative possibilities, we elect to pursue the one that seems to hold out the greatest promise for human betterment. A rigid adherence to non-maleficence in embryo protection or human protection would shut down stem cell research before we could even know if future wholeness by this means is boom or bust.

We believe it is ethical to gamble on stem cells. At present, scientists tell us that research needs to continue on embryonic stem cells, even though other possible methods of derivation of pluripotent stem cells may be on the horizon. To be honest, we may lose the bet. Still, the lure of good stewardship of dynamic science holds up before us a vision of a human race that is healthier, living longer, and enjoying a more abundant life. We believe we should say "yes" to such a possibility.

Notes

1. Cynthia Fox, *Cell of Cells: The Global Race to Capture and Control the Stem Cell* (New York: W. W. Norton, 2007), 59.

2. Francis Collins, *The Language of God: A Scientist Presents Evidence for Belief* (New York: Free Press, 2006), 256.

Index

About the Authors

Ted Peters is professor of systematic theology at Pacific Lutheran Theological Seminary, and is also a member seminary of the Graduate Theological Union. He is an ordained pastor in the Evangelical Lutheran Church of America. He received a Ph.D. from the University of Chicago and arrived in Berkeley, California, in 1978. Since 1982 he has participated in the expanding dialogue between natural science and Christian faith, working as a research scholar at the Center for Theology and the Natural Sciences. He served as principal investigator for the National Institutes of Health–funded research project "Theological and Ethical Questions Raised by the Human Genome Initiative," 1990–1994. He is a former editor of *Dialog, A Journal of Theology* and is now coeditor of *Theology and Science*. He has published widely on genetic research and its ethical implications. Since 2005 he has served as a medical ethicist on the Scientific and Medical Accountability Standards Working Group for the California Institute for Regenerative Medicine, otherwise known as Proposition 71.

Karen Lebacqz recently completed a three-decade career as Robert Gordon Sproul Professor of Theological Ethics at the Pacific School of Religion, a member seminary of the Graduate Theological Union. In 2005–2006, she served as bioethicist in residence at Yale University. She earned a Ph.D. from Harvard University. She has published widely on theories of justice, feminist ethics, professional ethics, and bioethics. Ordained in the United Church of Christ, she has served as consultant to the National Council of Churches, the World Council of Churches, the California State Department of Health, and to the U.S. Congress on policies dealing with human subjects in experimentation. Along with Ted Peters, she worked on the Center for Theology and the Natural Sciences project "Theological and Ethical Questions Raised by the Human Genome Initiative," funded by the National Institutes of Health. She chaired the Geron Corporation's Ethics Advisory Board when the isolation of human embryonic stem cells was first announced.

Gaymon Bennett is pursuing a Ph.D. at the Graduate Theological Union in the field of systematic theology with a special focus on ethics, anthropology, and politics. He began as a student of both Karen Lebacqz and Ted Peters, becoming a coteacher in courses dealing with genetics and ethics. He is a pastor in the Church of the Nazarene. With Ted Peters, he coedited the book *Bridging Science and Religion*, now translated from English into five other languages and read around the world. He currently serves as the director of ethics at the Synthetic Biology Engineering Research Center at the University of California, Berkeley.

Unknown to most outside observers, from the earliest days of embryonic stem cell research through today's latest developments, Christian theologians have been actively involved with leading laboratory research scientists to determine the ethical implications of stem cell research. Contrary to popular expectation, these religious ethicists have been courageously advocating in favor of research. Three of these dynamic theologians tell their story in *Sacred Cells? Why Christians Should Support Stem Cell Research.*

Sacred Cells? takes readers through the twists and turns of stem cell development, providing a brief history of the science and an overview of the competing ethical frameworks people use in approaching the heated debate. Each new scientific advance, from the cloning of Dolly the sheep to the use of engineered cells in humans, had to be carefully considered before proceeding. Rejecting the widely held belief that the ethics of stem cell research turn on the moral status of the embryo, the authors carefully weigh diverse ethical problems. Ultimately, they embrace stem cell research and the prospect of increased health and well-being it offers.